Proceedings of the 2nd International Conference on BioGeoSciences

Rolando Cárdenas · Vladimir Mochalov
Oscar Parra · Osmel Martin
Editors

Proceedings of the 2nd International Conference on BioGeoSciences

Modeling Natural Environments

Springer

Editors
Rolando Cárdenas
Universidad Central "Marta Abreu"
de Las Villas
Santa Clara, Cuba

Vladimir Mochalov
Institute for Cosmophysical Research
Kamchatka, Russia

Oscar Parra
University of Concepción
Concepción, Chile

Osmel Martin
Universidad Central "Marta Abreu"
de Las Villas
Santa Clara, Cuba

ISBN 978-3-030-04232-5 ISBN 978-3-030-04233-2 (eBook)
https://doi.org/10.1007/978-3-030-04233-2

Library of Congress Control Number: 2018961690

This Springer imprint is published by the registered company Springer Nature Switzerland AG
The registered company address is: Gewerbestrasse 11, 6330 Cham, Switzerland

Contents

On the Quantification of Habitability: Current Approaches

Rolando Cárdenas, Rosmery Nodarse-Zulueta, Noel Perez, Daile Avila-Alonso and Osmel Martin

Abstract In this chapter, we outline general ideas to quantify habitability, starting with a general abiogenesis–biogenesis conceptual model. We connect this model with the approach of the astrobiological school of quantitative habitability, specifically with quantitative habitability theory, to devise habitability indexes. We present two indexes devised by us: the Aquatic Primary Habitability for photosynthesis-based ecosystems, and the Chemosynthetic Habitability Index for chemoautotrophy-based ones. As a case study, we present the application of the last one to hydrothermal vents. It is also mentioned the possibility of embedding parameters such as net primary productivity, calculated using habitability indexes, into greater ecological models with several trophic models, making a clear connection between the astrobiological and ecological approaches of quantitative habitability.

Keywords Abiogenesis–biogenesis model · Quantitative habitability theory
Habitability index

1 Introduction

Quantitative habitability remains an open issue. Currently, there are three (complementary) approaches to quantify habitability. In the astrobiological, the main premises for life origin (abiogenesis) and evolution (biogenesis) are studied. In the biogeochemical, emphasis is put on biogeochemical cycles and the availability of nutrients and energy, while in the ecological a closer look is put at the interactions between the species in the context of the ecosystem [1]. However, due to the complexity of life, the quantification of habitability is still a very open science. Thus, our aim in this paper is to launch some ideas on how to quantify habitability

R. Cárdenas (✉) · R. Nodarse-Zulueta · N. Perez · D. Avila-Alonso · O. Martin
Planetary Science Laboratory, Universidad Central "Marta Abreu" de Las Villas, Santa Clara, Cuba
e-mail: rcardenas@uclv.edu.cu

R. Cárdenas et al. (eds.), *Proceedings of the 2nd International Conference on BioGeoSciences*, https://doi.org/10.1007/978-3-030-04233-2_1

in the spirit of this II International Conference on BioGeoSciences (BG-17): from the small molecular–cellular scales to the enormous astrobiological–cosmological ones.

We work with the hypothesis that the current level of knowledge of mankind allows generalizations concerning life in the Universe. So, accepting the principle of world's material unity (the most basic laws of Nature, formulated by physics and chemistry, are valid in the whole observed Universe, which is supported by a wide set of astrophysical observations), I propose a conceptual model for abiogenesis (life origin)–biogenesis (life evolution), in principle valid for the entire observed Universe. From [2, 3] the main ingredients of this model can be inferred, i.e., life needs:

(1) Biogenic chemical elements (for instance, CHON in Earth): the chemical (mineral) aspect of life;
(2) A liquid medium (solvent) for the biogenic elements properly mix to form biomolecules (water on Earth): the mixing (kinetic) aspect of life;
(3) An energy source for the above-mentioned mix proceeds at a proper speed, to keep low entropy and to perform work (light for photosynthesis, redox chemical energy for chemosynthesis): the energetic (thermodynamic) aspect of life;
(4) An appropriate physicochemical environment allowing formed biological molecules to persist, for instance; adequate temperature range, radiational regime, etc., the physicochemical aspect of life.

Rocky planetary bodies (either planets or satellites) are the places in the Universe in which above-mentioned premises are more likely to be fulfilled. It is very difficult to infer when they first formed, some authors saying that around 1–2 Gy after the Big Bang were necessary, while others state that 10–17 million could have been enough [4].

2 The Quantification of Habitability

The main postulate of the (astrobiological) quantitative habitability theory [5] states that habitability indexes HI can be constructed as a product of environmental functions $f_i(\{x_j\})$ which depend on sets of environmental variables $\{x_j\}$:

$$\mathrm{HI} = \prod_{i=1}^{n} f_i(\{x_j\}) \tag{1}$$

Typically, an HI is normalized so that it ranges between 0 (dead environment) and 1 (optimum for life). A crucial fact is that a properly devised HI can be used to assess the net primary productivity NPP (rate at which non-living matter transforms into living matter) of an environment:

$$NPP = HI \cdot NPP_{max} \quad (2)$$

where NPP_{max} is the maximum NPP.

Relating the main postulate of quantitative habitability theory with the conceptual model for abiogenesis–biogenesis, a generic habitability index can be written as:

$$HI = f_M f_K f_E f_{PC} \quad (3)$$

where f_M, f_K, f_E, and f_{PC} are functions representing the mineral, kinetic, energetic, and physicochemical aspects of life, respectively. For each of these aspects, we have used several functions, some devised by other authors, some devised by us. However, in order to build an HI, as in every model of Nature, we just include the functions of those variables which limit habitability; otherwise, the model could be untreatable from the mathematical point of view. For instance, an HI for a terrestrial ecosystem can include a function of the solvent water $f(H_2O)$, but in an aquatic ecosystem water is not limiting but optimum, which allows to disregard this function or, more properly, to set $f(H_2O) = 1$. Another issue is to consider the interactions (feedbacks) between variables.

2.1 *The Aquatic Primary Habitability Index*

Using the methodology of the former section, we investigated the (primary) habitability of aquatic ecosystems. It is usually acknowledged that the main environmental variables controlling life in these ecosystems are light, nutrients, temperature, and salinity. Then, we propose an HI for aquatic environments without salt stress, the Aquatic Primary Habitability (APH):

$$APH = f(L)f(N)f(T) \quad (4)$$

where $f(L)$, $f(N)$, and $f(T)$ are functions of light, limiting nutrient, and temperature, respectively. Particular versions of this index are:

$$APH_I = f(L)f(T) \quad (5)$$

and

$$APH_{II} = f(L)f(N) \quad (6)$$

The index represented in Eq. (5) was applied to environments where light, rather than nutrients, is a limiting variable [6]. Common examples are rocky planets and satellites orbiting red dwarfs, stars with a light emission much smaller than emission of solar-type stars.

The index represented by Eq. (6) was applied to Ana Maria Gulf, a Cuban gulf of importance for tourism and fisheries [7]. The function of light was inspired in the E model for photosynthesis [8]:

$$\frac{P(z)}{P_S} = \frac{1 - \exp(-E_{PAR}(z)/E_S)}{1 + E^*_{UV}(z)} \tag{7}$$

where $P(z)$ and P_S are the photosynthesis rates at depth z and the maximum possible, respectively. $E_{PAR}(z)$ and $E^*_{UV}(z)$ are the irradiances of photosynthetically active radiation (PAR) and ultraviolet radiation (UV) at depth z, respectively. Spectral UV irradiances are convolved with a biological action spectrum weighting the inhibitory effect on photosynthesis of each wavelength, which is indicated by the asterisk. E_S is a parameter of photosynthetic efficiency: the smaller its value, the greater the efficiency of the species to use PAR. Photosynthesis rates are depth-dependent, so the actual $f(L)$ we use in the aquatic habitability indexes presented above is an average in all the photic zone normalized respect to the optimum average:

$$f(L) = \left\langle \frac{P(z)}{P_S} \right\rangle / \left\langle \frac{P(z)}{P_S} \right\rangle_{opt} \tag{8}$$

The calculation of the optimum average was presented in Cardenas et al. [6]. The function on nutrients used in Eq. (6) was inspired in an eutrophication index [7, 9].

2.2 A Chemosynthetic Habitability Index

To devise a chemosynthetic habitability index (QHI), we propose to substitute the light function $f(L)$ in APH by a function of the chemical energy $f(Q)$, which chemoautotrophic organisms are able to use:

$$\mathrm{QHI} = f(Q)f(N)f(T) \tag{9}$$

The form of the chemical function $f(Q)$ can be inspired by suggested analogies between photosynthesis and chemosynthesis, as can be seen in Figs. 3 and 4 by Shock and Holland [1]. Ignoring the (still undetected) last stage of chemoinhibition in above-mentioned Fig. 4, we suggest, by analogy with the non-inhibitory part of the E model for photosynthesis, the following chemical function [10, 11]:

$$f(Q) = 1 - e^{-\frac{I}{q}} \tag{10}$$

where I is the chemical energy per unit area and per unit time that the chemoautotrophic organism receives, and q is a parameter of chemosynthetic efficiency.

The function of nutrients *f(N)* is inspired in the well-known Michaelis–Menten kinetics (normalizing respect to its maximum value):

$$f(N) = \frac{v_{\max}[\mathrm{LN}]/(K_{\mathrm{M}} + [\mathrm{LN}])}{f_{\max}(N)} \tag{11}$$

where [LN] stands for the concentration of limiting nutrient, $v_{\max}$ is the maximum assimilation rate of the limiting nutrient, and K_{M} is the semi-saturation constant known as Michaelis constant.

As a function of temperature *f(T)*, we take an inverted parabola symmetric respect to the optimum temperature for life T_{opt} [12]:

$$f(T) = 1 - \left(\frac{T_{\mathrm{opt}} - T}{T_{\mathrm{opt}} - 273}\right)^2 \tag{12}$$

3 Case Studies

We have applied habitability indexes to several scenarios, which can be divided into extraterrestrial, of Earth's geological history, current Earth and Cuban context. In Chap. 8 of these proceedings, the reader can find a detailed application of our Aquatic Primary Habitability to the Gulf of Ana Maria in Cuba, a photosynthesis-based ecosystem. Thus, in this chapter we present the application to a chemosynthesis-based ecosystem: the black smoker TY, located at 2.3 km depth in the East Pacific Ridge. Hydrothermal vents can host very dynamic ecosystems, mostly supported by chemosynthesis, although low levels of photosynthesis from geothermal photons have been reported by Perez et al. [13] and Das et al. [14]. Typically, the reaction which most contributes to chemosynthesis is the oxidation of hydrogen sulfide:

$$\mathrm{H_2S} + 2\mathrm{O_2} \rightarrow \mathrm{SO_4^{2-}} + 2\mathrm{H^+} \tag{13}$$

in which the limiting reactant is dioxygen. This is an extremely rich source of energy for chemoautotrophs, releasing a free energy:

$$\Delta G = 794\ \mathrm{kJ/mol} \tag{14}$$

To calculate *I*, we considered that the geometry of the hydrothermal fluid or effluent is a plane and that the reactants diffuse from it to both right and left directions in imaginary planes parallel to the hydrothermal fluid plane. Then, the Fick law was used to calculate the fluxes φ of dioxygen:

$$\varphi = -D\frac{d[O_2]}{dx} \tag{15}$$

where D is the diffusion coefficient of dioxygen, $[O_2]$ is its concentration, and x is the distance from the hydrothermal fluid plane. The intensity I is found from:

$$I = \varphi \Delta G \tag{16}$$

For the calculation of $f(Q)$, it was assumed a null concentration of dioxygen in the vent and a constant increase up to 5.0 mg/mol at 2.5 m from it. It was also considered the dependence on temperature of the diffusion coefficient D. For the calculation of the temperature function $f(T)$, it was assumed that the optimum temperature for living organisms is 298 K (25 °C), while for the calculation of the nutrients function $f(N)$ we considered the wide range of variation of nitrogen concentration $[N]$ in hydrothermal vents: 10, 100, and 1000 μmol/L. Then, the chemosynthetic habitability index (QHI) was calculated, and Figs. 1, 2, and 3 show the results.

From all three plots, we see that chemosynthetic habitability is highly sensitive to temperature and to the parameter of chemosynthetic efficiency q, while it shows little sensitivity to the concentration of nitrogen (limiting nutrient). However, it should be noticed that the value taken for the Michaelis constant K_M for the evaluation of the function of nutrients is one reported for a generic phytoplankton organism [15], given the scarcity of corresponding data for chemoautotrophs.

We have presented a first habitability index for chemosynthesis-based ecosystems, specifically for chemoautotrophy-based ones. The index has the spirit of quantitative habitability from an astrobiological perspective. Its mathematical form can be considered quite general, although some refinements can be done. Its application to a case study, hydrothermal vents, showed high sensitivity of chemoautotrophic life to temperature and to the parameter q, which characterizes the efficiency with which chemical energy is used by the chemoautotrophs. Scarcity

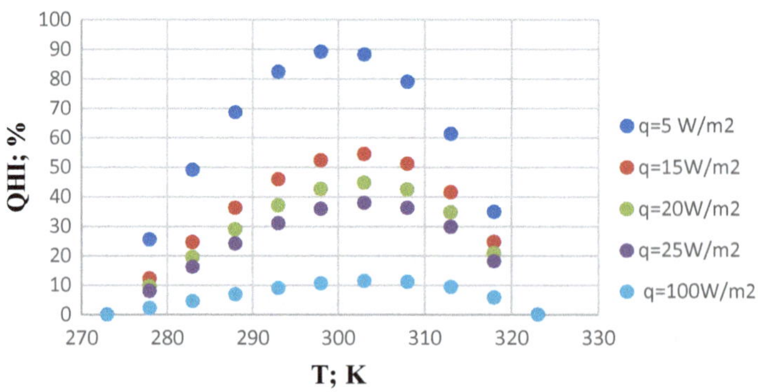

Fig. 1 Chemosynthetic habitability versus temperature for $[N]$ = 10 μmol/L

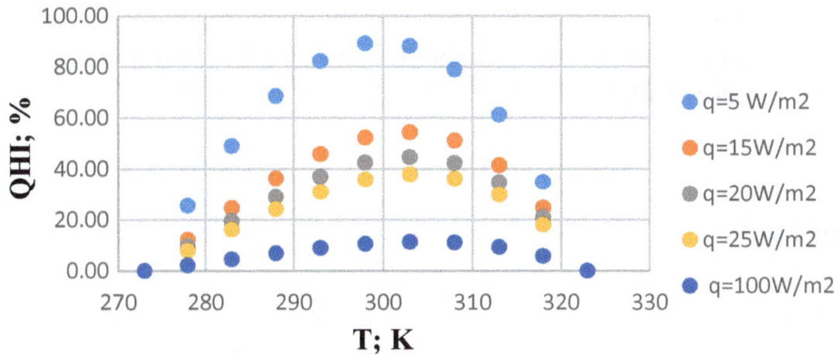

Fig. 2 Chemosynthetic habitability versus temperature for [*N*] = 100 μmol/L

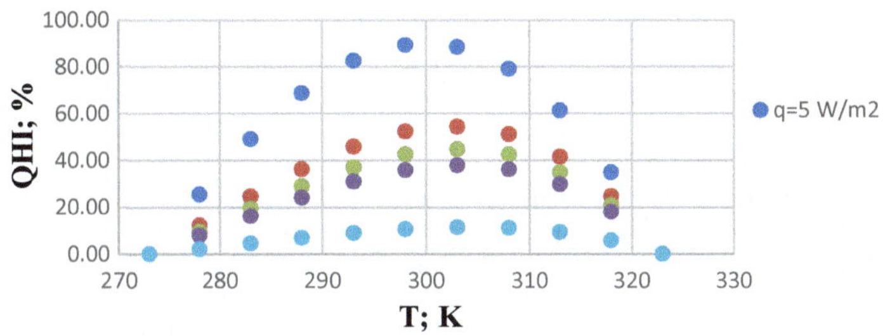

Fig. 3 Chemosynthetic habitability versus temperature for [*N*] = 1000 μmol/L

of data for these organisms still limits the refinement and applications of the index, especially in the deep biosphere. We hope that ongoing and planned expeditions involving deep sea and continental crust drilling will improve this situation.

On another hand, using Eq. (2) it is possible to estimate parameters, such as net primary productivity, for both photosynthesis- or chemosynthesis-based ecosystems. It is also possible to embed these parameters into greater ecological models, making a clear connection between the astrobiological and ecological approaches of quantitative habitability. In Chap. 8 of this book, an explicit example is presented.

4 Conclusions

In this chapter, we outlined a methodology to quantify habitability. It incorporates ideas from the astrobiological school, specifically quantitative habitability theory; to devise habitability indexes valid either for photosynthesis-based ecosystems or for chemoautotrophy-based ones. Applications to cases studies are presented or mentioned. It is also suggested to embed the parameters calculated using habitability indexes into ecological models of trophic levels. This also shows the applicability of quantitative habitability theory to spatial–temporal scales typical in ecological studies, showing a useful bridge between astrobiology and ecology. In forthcoming publications, refinements of this methodology will be presented, as well as its applications to case studies.

References

1. Shock E, Holland M (2007) Quantitative habitability. Astrobiology 7:839
2. Schulze-Makuch D, Irwin L (2008) Life in the Universe. Springer-Verlag, Berlin Heidelberg, Germany
3. Raven J, Cockell C, Kaltenegger L (2011) Energy sources for, and detectability of, life on extrasolar planets. Capítulo del libro genesis-in the beginning: precursors of life, chemical models and early biological evolution. Springer, Amsterdam
4. Loeb A (2014) The habitable epoch of the early universe. Int J Astrobiol 13:337
5. Méndez A (2010) Evolution of the global terrestrial habitability during the last century. Proceedings of sixth astrobiology science conference, Houston, TX, USA, pp 26–29
6. Cardenas R, Pérez-Díaz N, Martínez-Frias J, Martín-González O (2014) On the habitability of aquaplanets. Challenges 5:284
7. Álvarez-Salgueiro J (2015) Productividad primaria del fitoplancton en el golfo de Ana María, Cuba. B.Sc. Thesis in Physics. Supervisors: Dailé Avila-Alonso-Alonso & Rolando Cardenas. Consultant: Roberto González-de Zayas. Available at http://dspace.uclv.edu.cu:8089/
8. Fritz J, Neale P, Davis R, Pelloquin J (2008) Response of Antarctic phytoplankton to solar UVR exposure: inhibition and recovery of photosynthesis in coastal and pelagic assemblages. Mar Ecol Prog Ser 365:1
9. Karydis M, Ignatiades L, Moschopoulou N (1983) An index associated with nutrient eutrophication in the marine environment. Estuar Coast Shelf Sci 16:339
10. Cardenas R, Pérez Díaz N, Avila-Alonso D, Nodarse-Zulueta R (2017) ¿Se originó la vida en el eón Hadeico? ¿De manera fotosintética o quimiosintética? Memorias de la VII Convención de Ciencias de la Tierra, Abril/2017, La Habana, Cuba. Available at http://www.cubacienciasdelatierra.com/es/general15
11. Nodarse-Zulueta R (2017) Diseño de un Indice de Habitabilidad Quimiosintética. B.Sc. Thesis. Supervisors: Rolando Cardenas and Noel Pérez Díaz. Consultant: Dailé Avila-Alonso. Available at http://dspace.uclv.edu.cu:8089/
12. Volk T (1987) Feedbacks between weathering and atmospheric CO_2 over the last 100 million years. Am J Sci 287:763
13. Pérez N, Cardenas R, Martin O, Leiva-Mora M (2013) The potential for photosynthesis in hydrothermal vents: a new avenue for life in the Universe? Astrophys Space Sci 346:327–331
14. Das A, Singh T, LokaBharathi P, Dhakephalkar P, Mallik S, Kshirsagar P, Khadge N, Nagender Nath B, Bhattacharya S, Kumar Dagar A, Kaur P, Ray D, Shukla A, Fernandes C, Fernandes S, Thomas T, Mamatha S, Shashikant Mourya B, Murti Meena R (2017) Astrobiological implications of dim light phototrophy in deep-sea red clays. Life Sci Space Res 12:39–50
15. Amemiya T, Enomoto T, Rossberg A, Yamamoto T, Inamori Y, Itoh K (2007) Stability and dynamical behaviour in a real lake model and implications for regime shifts in real lakes. Ecol Model 206:54

The Dynamical Systems Approach to Modeling: The Universe as a Case Study

Ailier Rivero-Acosta, Adrian Linares-Rodriguez and Carlos R. Fadragas

Abstract The acceleration of the expansion of the Universe, as it is indicated by observations of redshift of light coming from supernovas, anisotropies of cosmological microwave background radiation, and the large-scale structure of the Universe, defines one of the most interesting theoretical problems that is facing the modern cosmology. The aim of this work is to show the analysis of the idea that inflation and dark energy are two subjects closely related, that is, both equivalents to the fundamental scalar field known as the standard model Higgs field. We considered that there exist non-trivial solutions with non-minimal coupling of the Cosmological Higgs Field to gravity. For this condition, an attractive cosmological model was derived. Results from applying the dynamical stability analysis show that the current accelerated expansion of the Universe is one of several possibilities. The future behavior of the Universe could seriously affect the existence of particles and structures that we are made of. For that reason, it is important to do some comments on this idea.

Keywords Cosmological Higgs Field · Dynamical systems · Accelerated expansion

1 Introduction

The acceleration of the expansion of the late Universe, as it is indicated by observations [1–6], defines one of the deepest theoretical problems that is facing modern cosmology. In July 2012 was announced the observation of the Higgs boson at the CERN. Then, it is a very stimulating task to build a cosmological model that includes the Cosmological Higgs Field non-minimally coupled to

A. Rivero-Acosta (✉) · A. Linares-Rodriguez · C. R. Fadragas
Universidad Central "Marta Abreu" de Las Villas, Santa Clara, Villa Clara, Cuba
e-mail: arfivero@uclv.cu

R. Cárdenas et al. (eds.), *Proceedings of the 2nd International Conference on BioGeoSciences*, https://doi.org/10.1007/978-3-030-04233-2_2

gravity and to study the viability of that model in a Friedmann–Robertson–Walker (FRW) Universe for describing the current acceleration of the expansion of the Universe.

The Cosmological Higgs Field (CHF) is expected to exist in our Universe. This field is related to the mechanism which creates inertial mass of particles known as the Brout–Englert–Higgs mechanism [7, 8].

In the last decades, the Higgs field has been used to explain the inflationary stage of the Universe for some authors [9, 10] and it is being considered as a possibility to describe the unknown agent called "dark energy" [11], which is assumed to be the responsibility of the current accelerated expansion of the Universe.

Although we are not going to focus on this direction, it is necessary to talk about another important aspect where the CHF has a great significance. As we said before, the CHF is related to the mechanism which gives mass to subatomic particles. If particles would not have mass, the gravitational interaction between them could not be possible. Therefore, the formation of structures in the Universe would not be possible, so the Universe as we know it today would not exist, and as a result, life would not exist. Another aspect that could also change radically the face of our world is the evolution of our Universe.

In the next section, we present the construction of our model. In Sect. 3, we show the dynamical analysis of the model. In Sect. 4, we discuss the results of our analysis. Finally, we conclude in Sect. 5 with some remarks.

2 Construction of the Model with the Higgs Field in a Cosmological Background

As we suggested in the previous section, the Cosmological Higgs Field (CHF) fills the entire Universe. In the present work, we represented the CHF as described by a singlet with a $U(1)$ symmetry which is given by:

$$\Phi(t) = \frac{1}{\sqrt{2}}\phi(t)e^{i\theta(t)} \tag{1}$$

where $\phi(t)$ and $\theta(t)$ are real-valued functions of time corresponding to the amplitude and phase of the CHF, respectively. Equations of the model are derived by the traditional way considering the Lagrangian formalism. The Friedmann equations read:

$$H^2 = \frac{k}{3}\left(\frac{1}{2}\dot{\phi}^2 + \frac{1}{2}\phi^2\dot{\theta}^2 + V(\phi) + \rho\right) \tag{2}$$

$$\dot{H} = k\left(-\frac{3\gamma - 2}{6}\rho - \frac{1}{3}\dot{\phi}^2 - \frac{1}{3}\phi^2\dot{\theta}^2 + \frac{V(\phi)}{3}\right) - H^2 \tag{3}$$

The Klein–Gordon equations, for amplitude function $\phi(t)$ and for phase function $\theta(t)$, are shown as follows:

$$\ddot{\phi} + 3H\dot{\phi} - \phi\dot{\theta}^2 + \frac{\partial V}{\partial \phi} = 0 \tag{4}$$

$$\ddot{\theta} + 3H\dot{\theta} - \frac{2\dot{\phi}\dot{\theta}}{\phi} = 0 \tag{5}$$

The energy conservation equation is written as follows:

$$\dot{\rho} + 3\gamma H\rho = 0 \tag{6}$$

The self-interaction potential $V(\phi)$ [7, 11] is written as follows:

$$V(\phi) = \frac{\lambda}{4}\phi^4 - \frac{\mu^2}{2}\phi^2 + \in \tag{7}$$

Next, we elaborated the model equations in order to realize the analysis of stability from the viewpoint of the dynamical system theory. Introducing the new variables $z = \dot{\phi}$ and $w = \dot{\theta}$, and taking $k = 1$, the dynamical Eqs. (2)–(6) of the model finally take the form:

$$\dot{z} = -3Hz + \phi w^2 - \lambda\phi^3 + \mu^2\phi \tag{8}$$

$$\dot{\phi} = z \tag{9}$$

$$\dot{w} = -3Hw - \frac{2zw}{\phi} \tag{10}$$

$$\dot{\rho} = -3\gamma H\rho \tag{11}$$

$$\dot{H} = -\frac{3\gamma - 2}{6}\rho - \frac{z^2}{3} - \frac{\phi^2 w^2}{3} + \frac{\lambda}{12}\phi^4 - \frac{\mu^2}{6}\phi^2 + \frac{\in}{3} - H^2 \tag{12}$$

3 Dynamical Analysis of the Model

Now, the dynamical system technique is applied to analyze the dynamics of the CHF and its cosmological implications (see [12–15]).

Equations obtained above (8)–(12) represent a system of first-order ordinary differential equations. Time variable does not appear explicitly at the right-hand side of these equations and then one says that they represent an autonomous dynamical system. The polynomial type of the self-interacting potential considered here leads to have no advantage by introducing the usual Hubble-normalized variables.

3.1 Determination of the Critical Points

The vector field (8)–(12) of the state space has the components $[z, \phi, w, \rho, H]$. The critical points can be found if the right-hand side of each equation of the system equals zero, and they are given in Table 1.

Table 1 Critical points of the equations system and their existence

P_i	z	ϕ	w	ρ	H	Existence
P_1	0	ϕ	$-\sqrt{-\mu^2+\lambda\phi^2}$	$\frac{4\in+2\mu^2\phi^2-3\lambda\phi^4}{-4+6\gamma}$	0	$\lambda\phi^2 \geq \mu^2$ and $\gamma \neq \frac{2}{3}$
P_2	0	ϕ	$\sqrt{-\mu^2+\lambda\phi^2}$	$\frac{4\in+2\mu^2\phi^2-3\lambda\phi^4}{-4+6\gamma}$	0	$\lambda\phi^2 \geq \mu^2$ and $\gamma \neq \frac{2}{3}$
P_3	0	$-\frac{\mu}{\sqrt{\lambda}}$	0	$\frac{-4\in\lambda+\mu^4}{4\lambda-6\gamma\lambda}$	0	$\gamma \neq \frac{2}{3}$
P_4	0	$-\frac{\mu}{\sqrt{\lambda}}$	0	0	$-\frac{\sqrt{4\in\lambda-\mu^4}}{2\sqrt{3}\sqrt{\lambda}}$	$4\in\lambda > \mu^4$
P_5	0	$-\frac{\mu}{\sqrt{\lambda}}$	0	0	$\frac{\sqrt{4\in\lambda-\mu^4}}{2\sqrt{3}\sqrt{\lambda}}$	$4\in\lambda > \mu^4$
P_6	0	$\frac{\mu}{\sqrt{\lambda}}$	0	$\frac{-4\in\lambda+\mu^4}{4\lambda-6\gamma\lambda}$	0	$\gamma \neq \frac{2}{3}$
P_7	0	$\frac{\mu}{\sqrt{\lambda}}$	0	0	$-\frac{\sqrt{4\in\lambda-\mu^4}}{2\sqrt{3}\sqrt{\lambda}}$	$4\in\lambda > \mu^4$
P_8	0	$\frac{\mu}{\sqrt{\lambda}}$	0	0	$\frac{\sqrt{4\in\lambda-\mu^4}}{2\sqrt{3}\sqrt{\lambda}}$	$4\in\lambda > \mu^4$
P_9	0	$-\frac{\sqrt{\frac{\mu^2-\sqrt{\mu^4+12\in\lambda}}{\lambda}}}{\sqrt{3}}$	$-\frac{\sqrt{2\mu^2-\sqrt{\mu^4+12\in\lambda}}}{\sqrt{3}}$	0	0	Not real
P_{10}	0	$-\frac{\sqrt{\frac{\mu^2-\sqrt{\mu^4+12\in\lambda}}{\lambda}}}{\sqrt{3}}$	$\frac{\sqrt{2\mu^2-\sqrt{\mu^4+12\in\lambda}}}{\sqrt{3}}$	0	0	Not real
P_{11}	0	$\frac{\sqrt{\frac{\mu^2-\sqrt{\mu^4+12\in\lambda}}{\lambda}}}{\sqrt{3}}$	$-\frac{\sqrt{2\mu^2-\sqrt{\mu^4+12\in\lambda}}}{\sqrt{3}}$	0	0	Not real
P_{12}	0	$\frac{\sqrt{\frac{\mu^2-\sqrt{\mu^4+12\in\lambda}}{\lambda}}}{\sqrt{3}}$	$\frac{\sqrt{2\mu^2-\sqrt{\mu^4+12\in\lambda}}}{\sqrt{3}}$	0	0	Not real
P_{13}	0	$-\frac{\sqrt{\frac{\mu^2+\sqrt{\mu^4+12\in\lambda}}{\lambda}}}{\sqrt{3}}$	$-\frac{\sqrt{\sqrt{\mu^4+12\in\lambda}-2\mu^2}}{\sqrt{3}}$	0	0	$\sqrt{\mu^4+12\in\lambda} > 2\mu^2$
P_{14}	0	$-\frac{\sqrt{\frac{\mu^2+\sqrt{\mu^4+12\in\lambda}}{\lambda}}}{\sqrt{3}}$	$\frac{\sqrt{\sqrt{\mu^4+12\in\lambda}-2\mu^2}}{\sqrt{3}}$	0	0	$\sqrt{\mu^4+12\in\lambda} > 2\mu^2$
P_{15}	0	$\frac{\sqrt{\frac{\mu^2+\sqrt{\mu^4+12\in\lambda}}{\lambda}}}{\sqrt{3}}$	$-\frac{\sqrt{\sqrt{\mu^4+12\in\lambda}-2\mu^2}}{\sqrt{3}}$	0	0	$\sqrt{\mu^4+12\in\lambda} > 2\mu^2$
P_{16}	0	$\frac{\sqrt{\frac{\mu^2+\sqrt{\mu^4+12\in\lambda}}{\lambda}}}{\sqrt{3}}$	$\frac{\sqrt{\sqrt{\mu^4+12\in\lambda}-2\mu^2}}{\sqrt{3}}$	0	0	$\sqrt{\mu^4+12\in\lambda} > 2\mu^2$

3.2 Determination of the Eigenvalues for Each Critical Point

To be able to make the stability analysis of each critical point, one needs to determine the eigenvalues of the matrix of the linearization of the system (8)–(12) and evaluate the matrix at that critical point. The procedure is as follows. The system (8)–(12) must be moved to each critical point, at each time. Next, applying the perturbation technique around each critical point, we can investigate the behavior of the system around that critical point. After the origin of the differential equations is moved to the critical point considered, the linearization method is applied, and the corresponding Jacobian matrix is transformed to the Jordan canonical form. The procedure above must be applied to each critical point. In each case, the set of five eigenvalues corresponding to a given critical point is determined and then we will try to describe the behavior of the dynamics around that critical point. It will be made in the following step. In Table 2, the set of five eigenvalues for eight of the critical points is shown.

The nature of these eigenvalues determines the behavior of the dynamical system near the critical point considered. When the critical point is hyperbolic, we can apply the Hartman–Grobman theorem (see [4]).

3.3 Small Description of the Critical Points

The system (8)–(12) exhibits the set of critical points labeled P_1, P_2, …, P_{16}. The critical points P_1, P_2, P_3, P_6, P_{13}, P_{14}, P_{15}, and P_{16} have a null value for the parameter H corresponding this situation to a static Universe. The points P_4 and P_7 exhibit a negative value for the coordinate H corresponding this case to a contracting Universe. On the other hand, critical points P_5, and P_8 exhibit a positive value for the coordinate H corresponding to an expanding Universe. For all the critical points given in Table 1, the value of coordinate z always equals zero which indicates that the "translational" kinetic energy is zero.

The coordinate ϕ can take different values, and these values may be of four types (see Table 1): a negative value for the critical points P_3, P_4, and P_5, a positive value for the critical points P_6, P_7, and P_8, a different negative value for the points P_{13} and P_{14}, and a different positive value for the points P_{15} and P_{16}. Two values for coordinate ϕ correspond to points in the graphical behavior of the self-interacting potential $V(\phi)$ versus ϕ where the function exhibits extreme values. For $\phi = \pm\frac{\mu}{\sqrt{\lambda}}$, the function $V(\phi)$ has a minimum. In Table 1 are displayed the location and existence conditions of these critical points, and $V(\phi)$ and some basic observables are given in Table 3.

The critical points P_4, P_5, P_7, and P_8 exist for the condition $4 \in \lambda \geq \mu^4$. This condition corresponds to a nonnegative value of the self-interacting potential $V(\phi)$; that is, $V(\phi)$ always will be nonnegative for these critical points. The points P_3 and

Table 2 Sets of eigenvalues for the critical points of the system

$\sum = \sqrt{-3\mu^4 - 32\lambda\mu^2 + 12 \in \lambda}$,

$\Xi_1 = \sqrt{\lambda^2\left(216\lambda^3 \in^3 - 54\lambda^2 \in^2 \Pi_1 - 6\lambda\mu^4 \in \left(3\lambda\mu^2 - 9\lambda\sqrt{\mu^4 + 12\lambda \in} + \mu^2\sqrt{\mu^4 + 12\lambda \in}\right) + \mu^8(\mu^2 - 12\lambda)\Pi_2\right)}$

$\Xi_2 = \sqrt{\mu^4 + 12\lambda \in} - 9\lambda\, \Xi_3 = \mu^2 + \sqrt{\mu^4 + 12\lambda \in}, \Pi_1 = \left(-27\lambda^2 - 8\lambda\mu^2 + \mu^4 + 2\lambda\sqrt{\mu^4 + 12\lambda \in}\right)$

$\Pi_2 = \left(\mu^2 + \sqrt{\mu^4 + 12\lambda \in}\right)$

P_i	λ_1	λ_2	λ_3	λ_4	λ_5
P_3	0	$-i\sqrt{2}\mu$	$i\sqrt{2}\mu$	$-\frac{\sqrt{\gamma}\sqrt{4\in\lambda-\mu^4}}{2\sqrt{\lambda}}$	$\frac{\sqrt{\gamma}\sqrt{4\in\lambda-\mu^4}}{2\sqrt{\lambda}}$
P_4	$\frac{\sqrt{4\in\lambda-\mu^4}}{\sqrt{3}\sqrt{\lambda}}$	$\frac{\sqrt{12\in\lambda-3\mu^4}}{2\sqrt{\lambda}}$	$\frac{\gamma\sqrt{12\in\lambda-3\mu^4}}{2\sqrt{\lambda}}$	$\frac{\sqrt{12\in\lambda-3\mu^4-\sum}}{4\sqrt{\lambda}}$	$\frac{\sqrt{12\in\lambda-3\mu^4+\sum}}{4\sqrt{\lambda}}$
P_5	$-\frac{\sqrt{12\in\lambda-3\mu^4}}{2\sqrt{\lambda}}$	$-\frac{\gamma\sqrt{12\in\lambda-3\mu^4}}{2\sqrt{\lambda}}$	$-\frac{\sqrt{4\in\lambda-\mu^4}}{\sqrt{3}\sqrt{\lambda}}$	$-\frac{\sqrt{12\in\lambda-3\mu^4+\sum}}{4\sqrt{\lambda}}$	$-\frac{\sqrt{12\in\lambda-3\mu^4+\sum}}{4\sqrt{\lambda}}$
P_6	0	$-i\sqrt{2}\mu$	$i\sqrt{2}\mu$	$-\frac{\sqrt{\gamma}\sqrt{4\in\lambda-\mu^4}}{2\sqrt{\lambda}}$	$\frac{\sqrt{\gamma}\sqrt{4\in\lambda-\mu^4}}{2\sqrt{\lambda}}$
P_7	$\frac{\sqrt{4\in\lambda-\mu^4}}{\sqrt{3}\sqrt{\lambda}}$	$\frac{\sqrt{12\in\lambda-3\mu^4}}{2\sqrt{\lambda}}$	$\frac{\gamma\sqrt{12\in\lambda-3\mu^4}}{2\sqrt{\lambda}}$	$\frac{\sqrt{12\in\lambda-3\mu^4-\sum}}{4\sqrt{\lambda}}$	$\frac{\sqrt{12\in\lambda-3\mu^4+\sum}}{4\sqrt{\lambda}}$
P_8	$-\frac{\sqrt{12\in\lambda-3\mu^4}}{2\sqrt{\lambda}}$	$-\frac{\gamma\sqrt{12\in\lambda-3\mu^4}}{2\sqrt{\lambda}}$	$-\frac{\sqrt{4\in\lambda-\mu^4}}{\sqrt{3}\sqrt{\lambda}}$	$-\frac{\sqrt{12\in\lambda-3\mu^4+\sum}}{4\sqrt{\lambda}}$	$-\frac{\sqrt{12\in\lambda-3\mu^4+\sum}}{4\sqrt{\lambda}}$
P_{13}	0	$\frac{-\sqrt{2}\sqrt{-\sqrt{2}\Xi_1+6\lambda^2\in\Xi_2-\lambda\mu^4\Xi_3}}{3\lambda\sqrt{\Xi_3}}$	$\frac{\sqrt{2}\sqrt{-\sqrt{2}\Xi_1+6\lambda^2\in\Xi_2-\lambda\mu^4\Xi_3}}{3\lambda\sqrt{\Xi_3}}$	$\frac{-\sqrt{2}\sqrt{\sqrt{2}\Xi_1+6\lambda^2\in\Xi_2-\lambda\mu^4\Xi_3}}{3\lambda\sqrt{\Xi_3}}$	$\frac{\sqrt{2}\sqrt{\sqrt{2}\Xi_1+6\lambda^2\in\Xi_2-\lambda\mu^4\Xi_3}}{3\lambda\sqrt{\Xi_3}}$
P_{14}	0	$\frac{-\sqrt{2}\sqrt{-\sqrt{2}\Xi_1+6\lambda^2\in\Xi_2-\lambda\mu^4\Xi_3}}{3\lambda\sqrt{\Xi_3}}$	$\frac{\sqrt{2}\sqrt{-\sqrt{2}\Xi_1+6\lambda^2\in\Xi_2-\lambda\mu^4\Xi_3}}{3\lambda\sqrt{\Xi_3}}$	$\frac{-\sqrt{2}\sqrt{\sqrt{2}\Xi_1+6\lambda^2\in\Xi_2-\lambda\mu^4\Xi_3}}{3\lambda\sqrt{\Xi_3}}$	$\frac{\sqrt{2}\sqrt{\sqrt{2}\Xi_1+6\lambda^2\in\Xi_2-\lambda\mu^4\Xi_3}}{3\lambda\sqrt{\Xi_3}}$
P_{15}	0	$\frac{-\sqrt{2}\sqrt{-\sqrt{2}\Xi_1+6\lambda^2\in\Xi_2-\lambda\mu^4\Xi_3}}{3\lambda\sqrt{\Xi_3}}$	$\frac{\sqrt{2}\sqrt{-\sqrt{2}\Xi_1+6\lambda^2\in\Xi_2-\lambda\mu^4\Xi_3}}{3\lambda\sqrt{\Xi_3}}$	$\frac{-\sqrt{2}\sqrt{\sqrt{2}\Xi_1+6\lambda^2\in\Xi_2-\lambda\mu^4\Xi_3}}{3\lambda\sqrt{\Xi_3}}$	$\frac{\sqrt{2}\sqrt{\sqrt{2}\Xi_1+6\lambda^2\in\Xi_2-\lambda\mu^4\Xi_3}}{3\lambda\sqrt{\Xi_3}}$
P_{16}	0	$\frac{-\sqrt{2}\sqrt{-\sqrt{2}\Xi_1+6\lambda^2\in\Xi_2-\lambda\mu^4\Xi_3}}{3\lambda\sqrt{\Xi_3}}$	$\frac{\sqrt{2}\sqrt{-\sqrt{2}\Xi_1+6\lambda^2\in\Xi_2-\lambda\mu^4\Xi_3}}{3\lambda\sqrt{\Xi_3}}$	$\frac{-\sqrt{2}\sqrt{\sqrt{2}\Xi_1+6\lambda^2\in\Xi_2-\lambda\mu^4\Xi_3}}{3\lambda\sqrt{\Xi_3}}$	$\frac{\sqrt{2}\sqrt{\sqrt{2}\Xi_1+6\lambda^2\in\Xi_2-\lambda\mu^4\Xi_3}}{3\lambda\sqrt{\Xi_3}}$

Table 3 Values of some cosmological parameters ω_ϕ and q and the potential $V(\phi)$

$\Xi = \mu^2 - \sqrt{\mu^4 + 12 \in \lambda}$

P_i	$V(\phi)$	ω_ϕ	q
P_3	$\in - \frac{\mu^4}{4\lambda}$	-1	Undefined
P_4	$\in - \frac{\mu^4}{4\lambda}$	-1	-1
P_5	$\in - \frac{\mu^4}{4\lambda}$	-1	-1
P_6	$\in - \frac{\mu^4}{4\lambda}$	-1	Undefined
P_7	$\in - \frac{\mu^4}{4\lambda}$	-1	-1
P_8	$\in - \frac{\mu^4}{4\lambda}$	-1	-1
P_{13}	$-\frac{\mu^2(\Xi)}{6\lambda} + \frac{(\Xi)^2}{36\lambda} + \in$	$\frac{9\mu^4 + 16\mu^2\sqrt{\mu^4 - 12\lambda\in} - 36\lambda\in}{12\lambda\in - 35\mu^4}$	Undefined
P_{14}	$-\frac{\mu^2(\Xi)}{6\lambda} + \frac{(\Xi)^2}{36\lambda} + \in$	$\frac{9\mu^4 + 16\mu^2\sqrt{\mu^4 - 12\lambda\in} - 36\lambda\in}{12\lambda\in - 35\mu^4}$	Undefined
P_{15}	$-\frac{\mu^2(\Xi)}{6\lambda} + \frac{(\Xi)^2}{36\lambda} + \in$	$\frac{9\mu^4 + 16\mu^2\sqrt{\mu^4 - 12\lambda\in} - 36\lambda\in}{12\lambda\in - 35\mu^4}$	Undefined
P_{16}	$-\frac{\mu^2(\Xi)}{6\lambda} + \frac{(\Xi)^2}{36\lambda} + \in$	$\frac{9\mu^4 + 16\mu^2\sqrt{\mu^4 - 12\lambda\in} - 36\lambda\in}{12\lambda\in - 35\mu^4}$	Undefined

P_6 exist for $\gamma \neq 2/3$. The points P_{13}, P_{14}, P_{15}, and P_{16} exist for $\sqrt{\mu^4 + 12 \in \lambda} \geq 2\mu^2$. The points P_9, P_{10}, P_{11}, and P_{12} do not exist in the real domain, so these ones are not important for us because they have no cosmological significance. For convenience, we will write $\in_0 = \mu^4/4\lambda$.

3.4 Stability Analysis of the Model

In this section, the analysis of the stability of the first-order perturbations of the equations system (8)–(12) near the critical points displayed in Table 1 is made. In order to determine the stability of the critical points of the system, it is necessary to obtain and analyze the eigenvalues of each one. These eigenvalues are given in Table 2. It is important also to point out that only the most significant points from the physical point of view were considered for the analysis.

The point P_4 is a solution dominated by the scalar field potential, with negative Hubble parameter H and ϕ. In this case, like ϕ represents the amplitude, it can be seen as a point with the phase $\theta = \pi$. This point corresponds with a past attractor, i.e., if we go back in time the field lines tend to go to the point, or in other words, is a source point of the field. We could get the above conclusion using the Hartman–Grobman theorem.

The point P_5 is a solution dominated by the scalar field potential, with a positive value of H and a negative value of ϕ. Just like in the point P_4, like ϕ represents the amplitude, it can be seen as a point with the phase $\theta = \pi$. This point corresponds with a late attractor, i.e., the field lines tend to go to the point, or in other words, is a sink point of the field. We could get to that conclusion using the Hartman–Grobman theorem.

The point P_7 is a solution dominated by the scalar field potential, with a negative value of H and a positive value of ϕ. This point corresponds, just like P_4, with a past attractor, i.e., if we go back in time the field lines tend to go to the point, or in other words, is a source point of the field. We could get to this conclusion using the Hartman–Grobman theorem.

The point P_8 is a solution dominated by the scalar field potential, with positive values of H and ϕ. This point corresponds, just like in the point P_5, with a late attractor, i.e., the field lines tend to go to the point, or in other words, is a sink point of the field. We could get to this conclusion using the Hartman–Grobman theorem. The points P_{13} and P_{14} have negative values of ϕ, just like for P_4 and P_5. In these cases, like ϕ represents the amplitude, it can be seen as a point with the phase $\theta = \pi$. The points P_{15} and P_{16} have positive values of ϕ. All the critical points $P_{13} - P_{16}$ have nonzero values of the term related to the "rotational" kinetic energy of the field.

3.5 Some Important Parameters

Besides the dynamical analysis of the model, we can extract information from some cosmological parameters like the equation of state parameter $\omega_\Phi = \frac{p_\Phi}{\rho_\Phi}$ and the deceleration parameter $q = -\left(1 + \frac{\dot{H}}{H^2}\right) = -\frac{a\ddot{a}}{\dot{a}^2}$. These parameters can be expressed as function of the new variables as follows:

$$\omega_\Phi = \frac{z^2 + \phi^2 w^2 - 2V}{z^2 + \phi^2 w^2 + 2V} \tag{13}$$

$$q = \frac{(3\gamma - 2)}{2}\frac{\rho}{3H^2} + 2\left(\frac{z^2}{6H^2} + \frac{\phi^2 w^2}{6H^2}\right) - \frac{V}{3H^2} \tag{14}$$

Values for those quantities above and for the potential are given in Table 3, for each critical point "considered there." It is a very interesting situation that for all the critical points of this system, the equation of state parameter of the scalar matter equals negative unity, $\omega = -1$, which corresponds to a Quintessence-type scalar field.

4 Discussion of Results

Having done the stability analysis and extracting the information given by the cosmological parameters presented above, we can discuss our results.

The points P_5 and P_8 correspond both to a de Sitter phase $(\omega_\phi = -1)$, i.e., the field has a repulsive character. These solutions are dominated by the scalar field potential, and they are consistent with a late accelerated expansion stage $(q = -1 < 0)$. These two solutions agree with the Universe we know until now.

There are another two points which have an interesting meaning. They describe a Universe in a contracting stage. This fact does not mean we should reject these solutions. The points which describe such a behavior are the points P_4 and P_7. These two points are past attractors, so they correspond each one of them to a contracting Universe. The parameter $\omega_\phi = -1 < 0$, indicating that the field has a repulsive character and because of $H < 0$, we have to interpret the sign of the deceleration parameter differently, giving as a result that the contraction is decelerated. This deceleration is caused by the effect of the scalar field.

5 Conclusions

In this work, we have investigated, from the perspective of the theory of dynamical systems, the dynamics of a cosmological model with the Cosmological Higgs Field (CHF) non-minimally coupled to gravity.

The introduction of the singlet structure to represent the Higgs cosmological field makes the corresponding dynamical model show a highly complex dynamic. Among the results provided by this cosmological model, once the dynamical analysis is carried out, it is that formally there can be universes both in expansion and in contraction, accelerated and decelerated.

We speculate that our Universe must be part of a dynamical multiverse-type system, with all possible dynamical behaviors, as expressed in the model.

As we said at the beginning of the article, we are going to do some comments about the relation between the behavior of the Universe and the possible future of life. If our Universe continues to expanding at an accelerated rate in some point of time, it will decay into a lower energy state. In this state, the kinds of particles we know today are not allowed to exist. On the other hand, if the Universe stops expanding and begins to contract, these particles would not exist either. Like we are made of these kinds of particles, these are bad news for us. This means that life could be something of not very long duration, compared to cosmological scales.

Apparently, the phase of the singlet structure does not play an important role in the dynamics of the Universe, but this must be reconsidered later.

Acknowledgements We want to thank R. Cárdenas for his useful comments and discussions. Furthermore, we thank Jose Monteagudo for information support. This research is partially supported by MES of Cuba.

References

1. Riess AG et al (1998) [Supernova Search Team Collaboration], Observational evidence from supernovae for an accelerating universe and a cosmological constant. Astron J 116:1009–1038. arXiv:astro-ph/9805201v1
2. Schmidt RW et al (2008) The High-Z supernova search: measuring cosmic deceleration and global curvature of the universe using type Ia supernovae. arXiv:astro-ph/9805200v1
3. Kowalski M et al (2008) Supernova cosmology project collaboration. Astrophys J 686(2):749. arXiv:0804.4142v1
4. Allen S et al (2008) Improved constraints on dark energy from Chandra X-ray observations of the largest relaxed galaxy clusters. Mon Not Roy Astron Soc 383(3):879–896. arXiv:astro-ph/0706.0033v3
5. Abazajian KN et al (2009) The seventh data release of the sloan digital sky survey. arXiv:astro-ph/0812.0649v2
6. Ade PAR et al (2014) (Planck Collaboration): Planck 2013 results. XXII. Constraints on inflation. arXiv:astro-ph/1303.5082v3
7. Nemiroff RJ, Patla B (2004) Decaying higgs fields and cosmological dark energy. arXiv:astro-ph/0409649v1
8. Englert F, Brout R (1964) Broken symmetry and the mass of gauge vector mesons. Phys Rev Lett 13:321
9. Bezrukov F, Shaposhnikov M (2007) The standard model higgs boson as the inflation. Phys Lett B
10. Fakir R, Unruh WG (1990) Improvement on cosmological chaotic inflation through nonminimal coupling. Phys Rev D
11. Rinaldi M (2015) Higgs dark energy. Class Quantum Gravity. arXiv:astro-ph/1404.0532v4
12. Leon G, Fadragas CR (2011) Cosmological dynamical systems and their applications. Lambert Academic Publishing, Sarrebruck
13. Coley AA (2003) Dynamical systems and cosmology. Dordrecht-Kluwer
14. Wiggins S (2003) Introduction to applied nonlinear dynamical systems and chaos. Berlin: Springer
15. Coley AA (1999) Dynamical systems in cosmology. arXiv:astro-ph/gr-qc/9910074v1

Enlarging Simple Ecological Models: Subspecies, Hidden Symmetries and Their Implications

Osmel Martin, Noel Perez-Diaz, Rolando Cárdenas and J. E. Horvath

Abstract Some basic principles to enlarge simple ecological models and the role of nonlinearities are discussed. The inclusion of internal groups and the new dynamic possibilities associated with this procedure are considered in the context of the logistic model. According to our results, processes like the success or extinction of a particular group without affecting the global population are not necessarily linked to the impact of environmental changes or the supremacy of a determined group or subspecies. In our case, the uniformity, the success or extinction of a particular group into a global population may be seen as the possibility to achieve or not a typical symmetry-breaking process. Such possibilities arise associated with the degree of nonlinearity contributions and the specificities of the interaction network in the model. Other elements linked with the ecological interaction, the role of symmetries and the phenomenological nature of ecological modelling are also discussed.

Keywords Ecological modelling · Subspecies · Symmetry · Symmetry-breaking process

O. Martin (✉) · N. Perez-Diaz · R. Cárdenas
Central University Marta Abreu from Las Villas, Santa Clara, VC, Cuba
e-mail: omgonzalez@uclv.cu

N. Perez-Diaz
e-mail: noelpd@uclv.edu.cu

R. Cárdenas
e-mail: rcardenas@uclv.edu.cu

J. E. Horvath
Departamento de Astronomía, Instituto de Astronomía, Geofísica y Ciencias Atmosféricas (IAG), USP, Sao Paulo, Brazil
e-mail: foton@iag.usp.br

R. Cárdenas et al. (eds.), *Proceedings of the 2nd International Conference on BioGeoSciences*, https://doi.org/10.1007/978-3-030-04233-2_3

1 Introduction

In the last decades, ecological modelling has emerged as a highly active field of research supporting in many cases economical activities and environmental decision-making [1–3]. In principle, ecological models try to express in a quantitative way the evolution and complexities associated with biological systems in their continuous interaction with the environment. Usually, with the increases of the field data and a better comprehension of the biological interaction network, preliminary models are improved or substituted by more accurate ones.

However, this is not necessarily always true in all cases. Ecological models respond to a determined spatial–temporal scale, and the amount of information included in the model varies according to the interests of the modellers and the purpose for which the model was conceived. For instance, the information given by a determined model could be considered as incomplete or limited in one context but enough in another. In most of the cases, the spirit and motivations of modelling at different scales are essentially different. At large scale, the modelling is more linked with the use of integrated magnitudes such as the biomass production or global populations of a determined species. Usually, this kind of modelling is extremely useful combined with environmental or climatic studies in the context of the world global change [4]. At smaller scale, the information tends to be more detailed, including, for instances, the spatial distribution or the interactions between different subspecies or population groups present in the global population [5].

On the other hand, more detailed models not necessary involve smaller spatiotemporal scales. In principle, the same ecosystem could be studied from both a global or integral vision and a more detailed one. For instance, to study the trophic interaction network in an aquatic environment the modeller declares only three classes or trophic levels such as phytoplankton, zooplankton and fish even when these classes are usually compounded by different subspecies or at least, they are separable by some classification criterion (stumps, colour, size and so on). It is obvious that these inclusions increase in a noticeable way the complexity of the model. However, it is also probable that both descriptions do not coincide at all, in the sense that the global magnitudes derived from the detailed model differ appreciably from the estimations due to the global one. This could be a serious problem if the estimations due to the original model agree with the observations.

During this work, in particular, we explore some underlying effects associated with the inclusions of new subspecies or populations groups in an already established ecological model.

2 Defining the Global and Detailed Models

As an illustrative example of the potential different between global and detailed description in the context of ecological modelling, we built in a rigorous way two different mathematical models (Global and Detailed) connected by a conservation law of form $X = \sum_1^N x_i$, where X is the total population, N the number of groups or partitions considered in the analysis and x_i the population belonging to the ith group.

2.1 *The Global Model*

With the aim to define the behaviour the behaviour of the population at global scale, let us suppose that in a local region the population of some hypothetical species X follows accurately this version of the logistic model:

$$\frac{\mathrm{d}X}{\mathrm{d}t} = M + rX\left(1 - \frac{X}{K}\right) \tag{1}$$

where M is a constant migration rate, r and k are, respectively, positive parameters representing the linear reproduction rate and the available sustaining resources of the environment (carrying capacity).

Introducing a new nondimensional variable for the modelled species of the form $Y = \frac{X}{K}$ and a scaled time $\mathrm{d}\tau = r\mathrm{d}t$, it is possible to rewrite the model (Eq. 1) in a new equivalent and simplified form:

$$\frac{\mathrm{d}Y}{\mathrm{d}\tau} = \tilde{M} + Y(1 - Y) \tag{2}$$

where the new form of the migration term is given by $\tilde{M} = \frac{M}{rK}$. Taking into account the dynamical equivalence between them and the obvious advantage of this new formulation, throughout this work, we are going to consider the system in the form of (Eq. 2) as our starting point to describe the dynamics of the population at global scale (hereinafter, the global model).

It is relatively easy to show that, according to Eq. (2), the global population always tends to the same asymptotic value given by $Y_{\text{Asymp}} = \frac{1}{2}\left(1 + \sqrt{1 + 4\tilde{M}}\right)$, regardless of the initial conditions. It is also clear that the above condition holds only when the migration parameter $\tilde{M}$ is constrained to be $\left(\tilde{M} \geq -\frac{1}{4}\right)$. For lower values, the solution of the system always diverges with time.

2.2 The Detailed Models

On the other hand, with the purpose to describe the behaviour of the different subspecies within the population, let us introduce the following family of models (hereinafter, detailed model), where for the sake of simplicity, only two different subspecies (y_1, y_2) are considered.

$$\frac{dy_1}{dt} = M\frac{y_1^n}{y_1^n + y_2^n} + y_1(1 - y_1 - y_2) \tag{3}$$

$$\frac{dy_2}{dt} = M\frac{y_2^n}{y_1^n + y_2^n} + y_2(1 - y_2 - y_1) \tag{4}$$

Now, the migration rate for each subspecies is modelled by a new parameter n. Some well-known representative cases are the following: when $n = 0$, migration rates are equal and constant $\left(\frac{M}{2}\right)$ for both subspecies, while for any other value (for instance $n = 0$ or $n = 1$) the condition of constant migration rate is relaxed to be some function (nonlinear in general) of the levels of individuals of the same group or subspecies into the global population as shown in Fig. 1.

Even when the structure of the migration rate considered here seems unnatural, the term has remarkable advantages. On one hand, the term mimic typical functional responses widely used in ecology (for instance, the Holling functions of the types I and II when $n = 1, n = 2$, respectively); on the other hand, its mathematical structure is more flexible than the original Holling functions to account the existence of subspecies.

On the other hand, with the purpose of describing the evolution of our two modelled groups, we introduce a *potential* of the form $V(A) = -\int \frac{dA}{dt} dA$ with the asymptotic condition $Y \rightarrow Y_{\text{Asymp}}$ for long enough times (see [6] for a similar treatment). Such procedure enables us to give a neat description of the stability features of the studied system, where now the minima and maxima of the potential function are associated with stability and instability conditions.

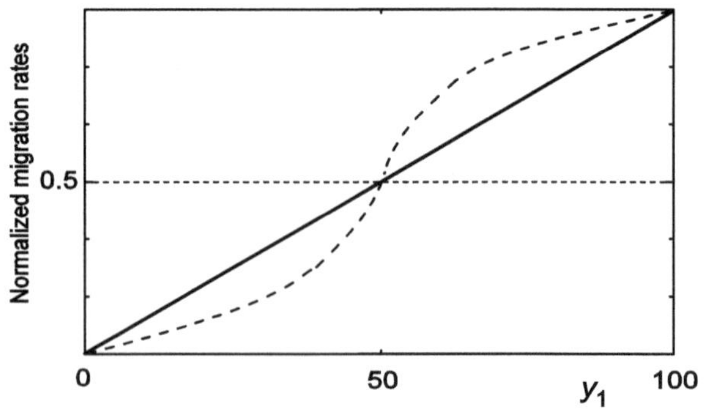

Fig. 1 Normalized migration rates laws of subspecies y_1 as a function of the relative population for each considered case. It is assumed that the total population is normalized as $y_1 + y_2 = 100$

During this work, we study the behaviour of three different potentials according to the values of the parameter n considered in our analyses.

$$V(A) = \frac{1}{2}A^2\left(Y_{\text{Asymp}} - 1\right) \quad n = 0 \tag{5}$$

$$V(A) = 0 \quad n = 1 \tag{6}$$

$$V(A) = \frac{1}{2}A^2\left(Y_{\text{Asymp}} - 1\right) - \tilde{M}Y_{\text{Asymp}} \ln\left[A^2 + Y_{\text{Asymp}}^2\right] \quad n = 2 \tag{7}$$

Let us explore how these potentials behave according to the specific sign of the migration parameter $\tilde{M}$. The results of these procedures are discussed in the next section.

3 Results and General Discussion

3.1 *A Constant Migration Rate* (n = 0)

According to the information displayed in Fig. 2, the detailed model with a constant migration rate reaches a unique stable configuration (only one minimum) when the migration parameter is positive. Under external fluctuations, the system always returns to a steady state where both subspecies have the same populations. Note that the solutions, in this case, share the same symmetry that the original system. Also, observe that the condition of neutral equilibrium $(\tilde{M} = 0)$ is simply translated in

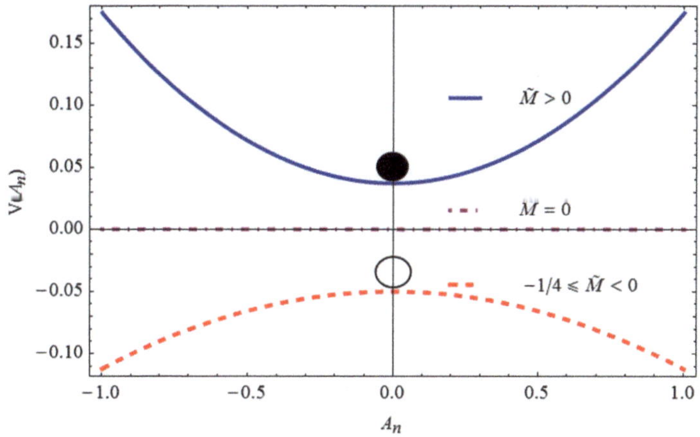

Fig. 2 Typical structures of the potential for the detailed model with a constant migration rate according to the sign of the parameter $\tilde{M}$

terms of the kinetic potential to a flat line indicating that there are not favoured compositions into the global population.

On the other hand, when the migration parameter is negative, the system has not steady states. Even when the population with equal composition is possible, this configuration is unstable against the effects of small fluctuations. In this case, both subspecies are unable to coexist in the same region. Driven by the potential, the asymmetry between both groups increases to reach its maximum value $A_n = \pm 1$, after that, the model breakdown. In this particular case, the solutions do not share the same symmetry with the original system, and the symmetry is spontaneously broken. However, the behaviour of the model at this point is meaningless by the presence of unbounded negative solutions.

3.2 The Neutral Case ($n = 1$)

In this particular case, we have a flat potential independently of the value of the migration rate. A flat potential could be considered as an indicator of the neutral stability of the system. Any asymmetric in the initial populations among the two different subspecies considered during our analysis remains unchanged over time. Note that the behaviour in this particular case is the same displayed by the system when the migration parameter vanishes, regardless of the specific value of n.

3.3 The Possibility of Alternative States: Success and Extinction

Even when there some similarities with the first scenario when $n = 0$, in this particular case, an uniform population (equal composition) is only possible when the migration parameter is negative $\left(-\frac{1}{4} \leq \tilde{M} < 0\right)$ as a consequence of the additional term in the potential (see Eq. 7). In this case, the process of symmetry breaking only takes place when the migration parameter is assumed positive. If such condition holds, any (initial or induced) asymmetry between both subspecies populations is amplified by the structure of the double-well potential shown in Fig. 3. However, in this case the detailed model does not break down in the limit, just when $A_n = \pm 1$, opposed to the variant where the migration rate is assumed as a constant. In some way, the structure of the model allows a natural transition under which a piece (the population of one subspecies) replaces completely the whole (the structure of the global population).

In this particular case, the possibility of success and extinction of identical subspecies accounted by the potential (Eq. 2) could be also interpreted as a typical spontaneous symmetry-breaking process, and a mechanism that, although considered in the context of the biological systems [7–10], reaches its maximum

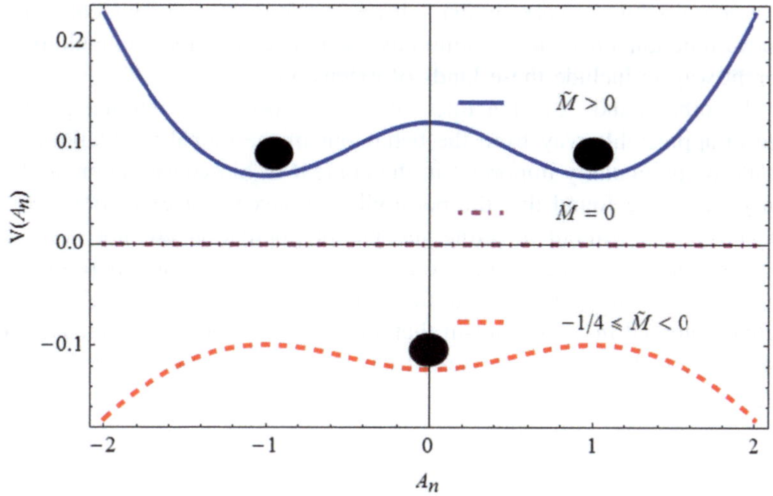

Fig. 3 Typical structures of the potential for the detailed model with a variable migration rate according to the sign of the parameter $\tilde{M}$

expression in the context of the theoretical physics [11, 12]. In the common usage, this term means that the system has a certain number of invariance properties that are not shared by its physical solution; hence, the solution has "spontaneously" broken the original symmetry. In our case, note that the detailed system is invariance when both subspecies are permuted simultaneously in the equations (Eqs. 3 and 4) as follows: $y_2 \rightarrow y_1; y_1 \rightarrow y_2$ while the obtained solutions are asymmetric.

Regrettably, despite the analogy with physical systems and the elegance of this method, the study of stability in terms of our potential is limited to considering only two different subspecies. More complex situations must be studied in the traditional form combining classical stability analyses and numerical experimentation.

4 Conclusions

Two basic questions have aimed at the development of this work. How to include the dynamic of subspecies or internal groups, and how do they behave when the global population follows a determined model? Probably, none of these questions have an immediate impact taking into account the high levels of uncertainties of the biological data, a situation that, undoubtedly, will change within a few years.

In principle, according to our results, the inclusion of different subspecies or populations groups in the context of a particular ecological model could be a rather problematic issue. Most of the difficulties associated with this procedure are linked with the high-nonlinear character of the ecological models [13, 14] and a lack of consensus within the ecological community about the own concept of population

[15]. Even in the case of very similar subspecies [16], the mathematical structure of the typical functional responses commonly used in ecology may difficult to adapt in a coherent way to include these kinds of extensions.

On the other hand, the dynamics of the subspecies or internal groups may diverge in appreciable way from the behaviour of the global population, an issue that could be particularly important in the context of bio-conservation studies [16, 17]. In general, we found that the possibility of success or extinction of a determined subspecies without affecting the global population are not necessarily a consequence of its intrinsic abilities to compete or deal with the environment. Such processes are possible even when the competing subspecies display the same ecological potential in a mechanism that shares a noticeable resemblance with the typical symmetry-breaking processes widely used in the context of field theories [7, 18, 19].

References

1. Schmolke A et al (2010) Ecological models supporting environmental decision making: a strategy for the future. Trends Ecol Evol 25(8):479–486
2. Grimm V et al (2014) Towards better modelling and decision support: documenting model development, testing, and analysis using TRACE. Ecol Model 280:129–139
3. Elith J, Leathwick JR (2009) Species distribution models: ecological explanation and prediction across space and time. Annu Rev Ecol Evol Syst 40(1):677–697
4. Pearson RG, Dawson TP (2003) Predicting the impacts of climate change on the distribution of species: are bioclimate envelope models useful? Glob Ecol Biogeogr 12(5):361–371
5. Kreuzer M, Tribsch A, Nyffeler R (2014) Ecological and genetic differentiation of two subspecies of Saussurea alpina in the Western Alps. Alp Bot 124(1):49–58
6. Gleiser M, Thorarinson J (2006) Prebiotic homochirality as a critical phenomenon. Orig Life Evol Biosph 36(5):501–505
7. Li R, Bowerman B (2010) Symmetry breaking in biology. Cold Spring Harb Perspect Biol 2 (3):a003475
8. Borile C et al (2012) Spontaneously broken neutral symmetry in an ecological system. Phys Rev Lett 109(3):038102
9. Sayama H, Kaufman L, Bar-Yam Y (2000) Symmetry breaking and coarsening in spatially distributed evolutionary processes including sexual reproduction and disruptive selection. Phys Rev E 62(5):7065–7069
10. Price RIA et al (2016) Symmetry breaking in mass-recruiting ants: extent of foraging biases depends on resource quality. Behav Ecol Sociobiol 70(11):1813–1820
11. Djouadi A (2008) The anatomy of electroweak symmetry breaking: Tome I: The Higgs boson in the Standard Model. Phys Rep 457(1–4):1–216
12. Gabrielli E et al (2014) Towards completing the standard model: vacuum stability, electroweak symmetry breaking, and dark matter. Phys Rev D 89(1):015017
13. Kratina P et al (2009) Functional responses modified by predator density. Oecologia 159 (2):425–433
14. Morozov AY (2010) Emergence of Holling type III zooplankton functional response: bringing together field evidence and mathematical modelling. J Theor Biol 265(1):45–54
15. Millstein RL (2009) Populations as individuals. Biol Theory 4(3):267–273
16. Patten MA (2015) Subspecies and the philosophy of science. Auk 132(2):481–485

17. Stein BA et al (2013) Preparing for and managing change: climate adaptation for biodiversity and ecosystems. Front Ecol Environ 11(9):502–510
18. Longo G, Montévil M (2011) From physics to biology by extending criticality and symmetry breakings. Prog Biophys Mol Biol 106(2):340–347
19. Goryachev AB, Leda M (2017) Many roads to symmetry breaking: molecular mechanisms and theoretical models of yeast cell polarity. Mol Biol Cell 28(3):370–380

Multi-agent Question-Answering System

Anastasia Mochalova and Vladimir Mochalov

Abstract The paper suggests an approach to the design of a question-answering system based on the interaction of various agents whose work is aimed at obtaining the answer most relevant to the user's request. The types of such agents, the principles of their work, their functions, the ways of interaction for obtaining the final answer are described. A distinctive feature of the described approach to the implementation of the multi-agent question-answering system is that among the agents providing the system operation, there are those using the machine data and the logical conclusions, as well as those whose main function is to communicate with people and receive the necessary information from them. Thus, the efficiency of such a system is largely determined by the fact that the system uses the most powerful intellectual resources—humans—along with the machine resources and algorithms. A peculiar feature of the question-answering system described in the paper is that the agents ensuring the system operation interact with each other, as well as with users in natural language.

Keywords Question-answering systems · Intelligent agents · Automatic text processing

1 Introduction

The paper considers an agent-oriented approach to the solution of the problem of question-answering system development.

A question-answering system (QAS) is an information system capable of receiving the user's natural language questions and of answering them in a natural language (see Fig. 1). The aim of development of such a system is to cut the time

A. Mochalova (✉) · V. Mochalov
Institute of Cosmophysical Research and Radio Wave Propagation FEB RAS, Mirnaya str. 7, Kamchatka Region, Elizovskiy, Paratunka 684034, Russia
e-mail: stark345@gmail.com

R. Cárdenas et al. (eds.), *Proceedings of the 2nd International Conference on BioGeoSciences*, https://doi.org/10.1007/978-3-030-04233-2_4

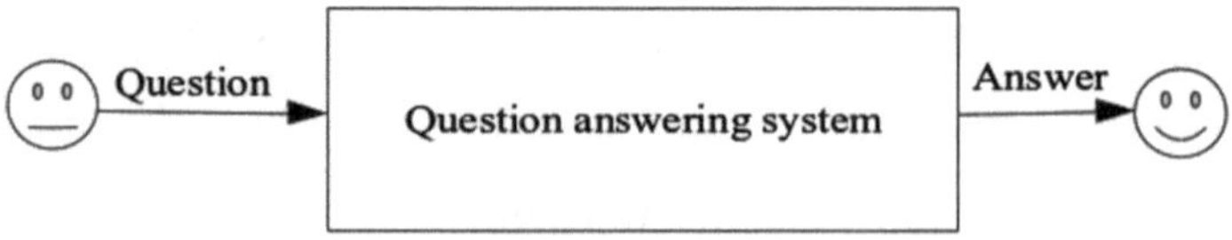

Fig. 1 Question-answering system

for search of some necessary information stored in large volumes of machine-readable texts in a natural language and to structure databases.

Creation of question-answering systems refers to the artificial intellect tasks. The topicality of development of the systems allowing a user to communicate with a computer in a natural language is determined by the constant increase of texts' volume represented in a machine form and by the necessity of a fast way of searching for desired information in these texts.

2 Types of Agents

The agents providing the question-answering system operation can be classified according to the following two types: machine-oriented agents and human-oriented agents (see Fig. 2).

The machine-oriented agents are a set of information systems and machine-readable texts which provide the required data for the question-answering system. In this case, machine-readable texts may store the data both in structured (dictionaries, thesauruses, ontologies, etc.) and in unstructured forms (different electronic documents represented in the form of texts of sites, books, journals, articles, etc.).

Human-oriented agents are the information systems providing communication with people to obtain the information required for question-answering system operation.

Among the main functions of the human-oriented agents, we can emphasize the following:

1. Search and realization of the communication with Internet users (solvers);
2. Analysis of users' messages on forums, Web sites;
3. Communication with people in a natural language;
4. Extraction of information about people (e.g., age, gender, country of residence, communication language, occupation, hobby, interests, aims of communication and problems);
5. Formation of databases of people whose knowledge can be applied in the question-answering system operation;

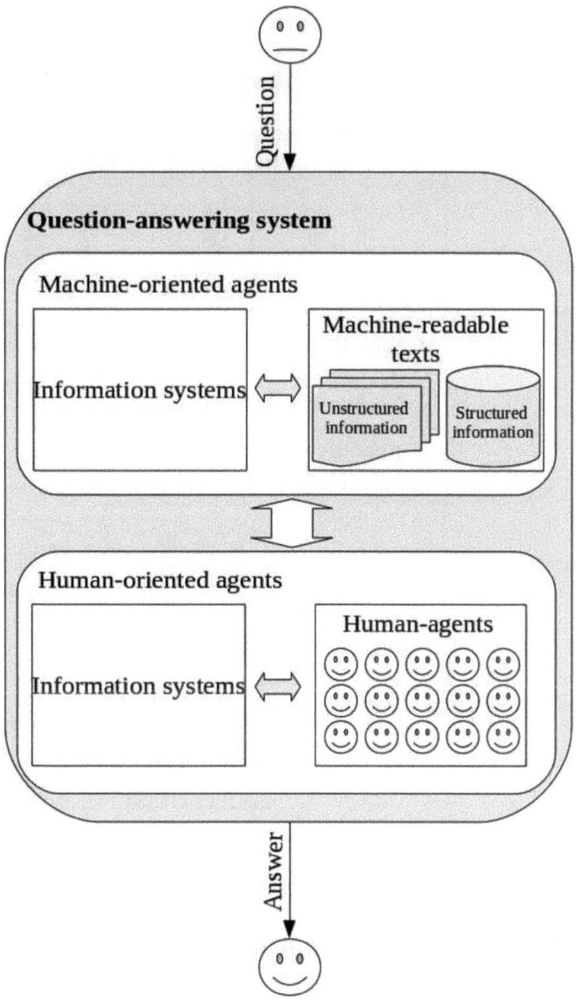

Fig. 2 Multi-agent collaborative question-answering system

6. Selection and attracting of people to solve a given task taking into consideration mutual interests;
7. Control of the dialogue between the information system and a human;
8. When necessary, checking the results.

Figure 3 schematically shows some different functions of the human-oriented agent–manager.

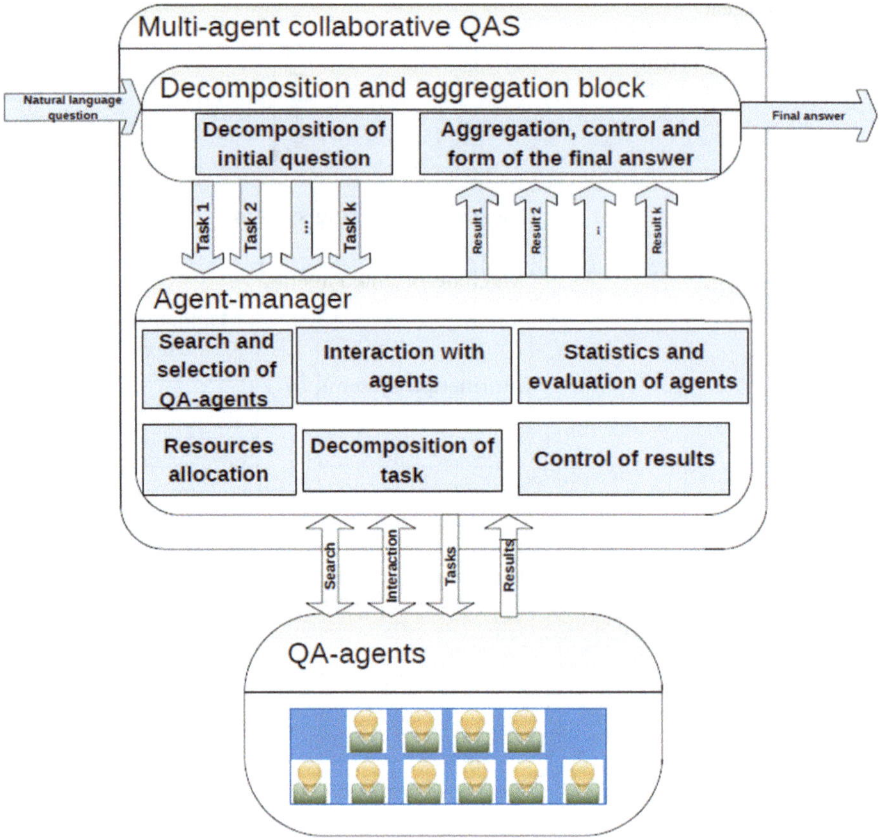

Fig. 3 Some different functions of the human-oriented agent–manager

3 Agent Interaction and Executing Tasks

During the process of question-answering system implementation, it is often reasonable to apply both agent types presented above to solve different subtasks some of which will be described below.

3.1 Question Classification

Question classification is an important component of any question-answering system. The operation of a question-answering system on the whole much depends on the correctness of determination of a question type.

The system parameters may affect the classification choice. They are: language, on which the system is oriented, its subject orientation, scales (volume of document collection and of different bases, dictionaries, ontologies, and other information repositories where we seek for an answer), and so on [1].

Conventionally, all questions may be divided into two global types:

1. Questions, on which an answer in the form of a fragment of an analyzed machine-readable text in a natural language can be given;
2. Questions, on which there is no answer in the direct form in the analyzed machine-readable texts (when a text fragment can be proposed as an answer); it is necessary to apply a logical deduction apparatus or additional information storage, structured or unstructured, to answer such questions.

At present time, there are many classifications of interrogative sentences. Different factors may affect the choice of classification, for example, a subject domain of a developing system, the language, in which a user will ask questions.

Lehnert, an author of one of the first papers [2] in which interrogative sentences were classified, distinguished 13 types of questions: causal antecedent, goal orientation, casual consequent, enablement, verification, disjunctive, instrumental/procedural, concept completion, expectation, judgmental, quantification, feature specification, and request. Later, Graesser in the paper [3] extended the classification suggested by Lehnert and added five more types of interrogative sentences: definition, example, interpretation, assertion, and comparison. As the question-answering systems developed, the question classification problems became more topical. Many developers of such kind of systems offered their unique classifications more suitable for the needs of the developed systems. Quite a large number of papers are devoted to the description of different classifications of interrogative sentences [4–9].

Sometimes, it is possible to determine the type of a question entering a system using only machine-oriented agents. However, a user often asks a question which cannot be adequately accepted without additional comments (e.g., when a person asks a question in the context of his/her thoughts).

It is clear that analysis of a user's profile and his/her request history can be realized by machine-oriented agents; however, it is not always enough for meaningful understanding of a question. Sometimes, some user's comments are necessary. At this stage, human-oriented agents are used. In the example under consideration, it is logical to consider the user asking a question a human-oriented agent; a question-answering system may ask a clarification question. In some cases, the system needs to ask a number of clarification questions.

It is clear that in case of uncertainty in a question, interaction of question-answering system with human-oriented agents is required.

3.2 *Question Decomposition*

The decomposition problem of a question entering the input of a question-answering system is reduced to the detection of several separate questions (in case if they exist). For example, the question "When and who discovered penicillin?" contains two different questions: "When was the penicillin discovered?" and "Who discovered the penicillin?". As a rule, the decomposition problem of a question in a question-answering system is solved using only machine-oriented agents. We should note that the decomposition module operation depends directly on a chosen system for question classification.

3.3 Clarification of Questions

Some user's questions addressed to a question-answering system need to be clarified in spite of their lexical ambiguity. For example, the question "How much does the ticket for 'Nutcracker' ballet cost?" evidently implies the following clarifying questions, without which a user cannot receive the correct answer: "In what town are you going to see the ballet?", "In what theater?", "What date would you like to visit the theater?", "What seats do you prefer?", and others. Only the author can answer these clarifying questions; thus, a question-answering system interacts with an agent–user at this stage.

In some cases, the clarifying questions can be formulated by a system without human-oriented agents. It occurs just in such a way in question-answering systems applying a solution tree to give an answer to a user.

3.4 Choice of Documents Containing an Answer

As a rule, a question-answering system seeks for an answer given by a user in structured information repositories (ontologies, dictionaries, databases) or in the collections of unstructured machine-readable texts. Such collections can be represented by either specialized libraries or universal libraries (e.g., a variety of Web resources of Internet).

There are different strategies for search of documents more relevant to a user's request. As a rule, a question-answering system scans the documents which are considered to be more relevant to the question. The number of documents does not exceed a defined quantity. For example, in the question-answering system QA-LaSIE [10], the system scans 20 documents defined as the most corresponding to a user's question by the results of search engine Okapi.

The choice of documents supposedly containing an answer is a problem which requires fast analysis of a large number of machine-readable texts in a natural language. Thus, at this stage of question-answering system operation, machine-oriented agents are primarily applied. They are capable of processing large volumes of data.

3.5 Creation of Structured Information Repositories

During the process of answering a question, most of the question-answering systems use information from structured information repositories: dictionaries, ontologies. Such repositories can be created either manually or in an automated way; that is, it requires human participation. Thus, to create structured information repositories, applied in question-answering systems, both machine-oriented (for automation of the process) and human-oriented agents (for putting together and enrichment of repositories manually) are used.

The authors of the paper [11] divide the automated data input into an ontology into two types:

1. Automated data input with a traditional lexicographical information (encyclopedical, definition, and other dictionaries and databases);
2. Automated data input based on the analysis of distributive characteristics of the lexicon in the text corpus.

One of the most widespread techniques of the automatic creation and enrichment of ontologies is based on the analysis of dictionary definitions (by template search in dictionary definitions). The idea of the automatic creation of ontologies based on dictionary article analysis is described in the papers [12–14].

An alternative approach to the automatic filling of ontologies is suggested in the paper [15] where the authors offer to modify the ontology by the semantic analyzer (simultaneously applying machine-oriented and human-oriented agents).

We should separately mention the necessity of application of human-oriented agents to construct answer databases which (e.g., such as in [16]) can be applied in question-answering systems. Some of such databases can be created on the basis of servers where people may ask a question, on which users may give arbitrary answers (e.g., [17]) or on the basis of forums.

3.6 Mapping Parts of Text with Ontology Nodes

A problem of word sense disambiguation is usually understood as a process of definition of an indivisible sense entity of a text, depending on the context where it has been used.

The need to determine the correspondence of parts of the text and ontology elements arises when solving a whole series of computer linguistics problems related to the automatic text processing (e.g., when implementing machine translation systems, automatic annotation and abstracting, development of information retrieval and question-answering systems, text corpus annotation systems).

When solving the problem of mapping parts of text with ontology nodes, one can define the following stages:

1. Preliminary processing of the text;
2. Definition of the boundaries of the sentences;
3. Allocation of syntaxemes boundaries;
4. Determination of possible lemmas variants for all allocated syntaxemes;
5. Search in the ontology of elements corresponding to the lemmas in stage 4;
6. Selection among the elements of the ontology found in stage 5 those that correspond to the syntaxemes of stage 3.

The first stage, "preliminary processing of the text," can include such actions on processing a text in natural language, presented in electronic form, as removal of text formatting symbols, removal of extra spaces and line breaks, fixing typos and all types of computer-recognizable errors in text arrangement.

For solving the second stage, the "definition of the boundaries of the sentences," there are many software implementations on the Internet that perform such division. However, the basis of most of these programs is the principle of determining the end of the sentence for the terminal punctuation mark (period, question, or exclamation point). Such approach to the problem of segmentation of sentences is attractive due to its simplicity, but in a real programming system of mapping parts of text with ontology nodes it is undesirable to employ it, because the number of erroneously found boundaries of the sentences using the described approach is unreasonably high.

The third stage, "allocation of syntaxemes boundaries," is definitely the most complex one of the tasks directly preceding solving the problem of mapping parts of text with the ontology nodes. By syntaxemes, we will mean a unit of text, which is defined in [18] as the minimal, further indivisible semantic-syntactic unit of the Russian language, which acts simultaneously as the carrier of the elementary (categorical-semantic) meaning and as a constructive element of more complex syntactic constructions. The correctness of allocation of syntaxemes boundaries in the analyzed text directly influences the quality of the system of mapping parts of this text to the ontology nodes.

In real examples having set the task of determining the mapping of parts of text with the elements of Wikidata (a database that can also be classified as an ontology), we will encounter difficulties in defining the boundaries of syntaxemes. Obviously, the correctness of the mapping of the parts of this sentence with the ontology elements directly depends on how the boundaries of the syntaxemes will be determined in the analyzed sentence. After the syntaxemes boundaries have been defined in the analyzed text, it is necessary to define lemmas (initial forms) for all syntaxemes, because the ontology elements are usually stored in the initial form. The next step is the search in the ontology for the elements corresponding to the lemmas. At the final stage, it is required to select a single one from the ontology elements found for all possible lemmas of each syntaxeme.

A brief review of methods and algorithms for solving word sense disambiguation is given in [19]. Among the approaches to solving word sense disambiguation, one can distinguish the methods based on the use of external sources of information and the methods based on machine learning (usually, semantically marked corpora are used for this). The author of [20] classifies the methods for solving word sense disambiguation by the type of external information sources used:

1. Structured data sources (machine-readable dictionaries, thesauruses, ontologies);
2. Non-structured data sources as text corpora, subdivided into:

   2.1. Non-marked corpora;
   2.2. Syntactically and/or semantically marked corpora.

One of the efficient approaches to solving the problem of mapping parts of text with ontology nodes is the use of rules that take into account the context in which the syntaxeme to be evaluated is used, and information from the ontologies, the structured sources of information.

We will assume that Steps 1–5 of Sect. 3.6 have already been completed, and we are working with a set of syntaxemes, each of which is associated with a set of the ontology elements. Then, the problem of word sense disambiguation reduces to selecting from each set of elements corresponding to a separate syntaxeme a single one that best reflects the lexical meaning of the syntaxeme under consideration.

When constructing the rules defining correspondence of the syntaxeme to an ontology element, one must take into account:

1. The context of the syntaxeme (the text closest to the syntaxeme is of the greatest importance: The most important for analysis is the text of the sentence in which the syntaxeme is used, followed by the text of the paragraph containing this sentence, then the section containing the sentence, then the section one level higher (e.g., chapter or paragraph), etc., ending with all analyzed text);
2. The semantic similarity of the syntaxeme, mapped during the analysis with the specific element of the ontology, to the syntaxemes from the context, taking into account the "proximity" of the context to the syntaxeme being analyzed. To determine the semantic similarity, one can use not only the ontology, but also the associative dictionaries;
3. Subject of the text. To determine the subject of the text, one can either ask the user to define it (e.g., by offering to choose from the list) or determine it automatically: At present, there are many algorithms for the automatic text classification.

3.7 *Search in Machine-Readable Texts in a Natural Language*

Question-answering systems, the aim of which is the answer on a user's question on a concrete text in a natural language, should naturally have a search system providing the search for desired information in a text. There are several kinds of such search systems. The most widespread is the full-text Web search engine (Google type). However, the systems capable of making a semantic search are becoming more popular. As a rule, crawlers are realized without direct participation of human-oriented agents due to the fact that a human cannot rapidly process large amounts of texts.

4 Conclusion

Development of question-answering systems is becoming a topical question nowadays. In the paper, we have showed the necessity of application of machine-oriented and human-oriented agents in the development of such systems. Functions of such agents have been described. The necessity to realize the interaction of agents of different types to provide qualitative operation of a question-answering system has been shown.

References

1. Mochalova AV (2017) The semantic analyzer of the Russian-language text for the question-answer system. Ph.D. dissertation, Petrozavodsk State University, 129 p
2. Lehnert W (1977) The process of question answering. Ph.D. dissertation, research report. No. 88. Yale University
3. Graesser A (1994) Question asking during tutoring. Am Educ Res J 31:104–137
4. Lapshin VA (2012) Question-answer systems: development and prospects. Sci Tech Inf Inf Process Syst 6:1–9
5. Cai L, Zhou G, Liu K et al (2011) Large-scale question classification in cQA by leveraging Wikipedia semantic knowledge. In: Proceedings of ACM CIKM, ACM, New York, pp 1321–1330
6. Huang Z, Thint M, Qin Z (2008) Question classification using head words and their hyperonims. EMNLP 927–936
7. Laokulrat N (2013) A survey on question classification techniques for question answering. KMITL Inf Technol J 2(1)
8. Roberts K, Masterton K, Fiszman M et al (2014) Annotating question types for consumer health questions. LREC workshop on building and evaluating resources for health and biomedical text processing
9. Sundblad H (2007) Question classification in question answering systems. Ph.D. dissertation, Linkopings University
10. Scott S, Gaizauskas R (2001) QA-LaSIE: a natural language question answering system. In: Proceedings of the 14th biennial conference of the Canadian society on computational studies of intelligence, pp 172–182
11. Rubashkin VS, Kapustin VA (2008) Use of definitions of terms in encyclopaedic dictionaries for automated ontology replenishment. In: XI all-Russian joint conference "internet and modern society", St. Petersburg
12. Hovy E, Knight K, Junk M Large resources. Ontologies (SENSUS) and Lexi-cons. www.isi.edu/natural-language/projects/ONTOLOGIES.html
13. Aramaki E, Imai T, Kashiwagi M, Kajino M, Miyo K, Ohe K (2005) Toward medical ontology using natural language processing. http://www.m.u-tokyo.ac.jp/medinfo/ont/paper/2005-aramaki-1.pdf

14. Rubashkin VS (2013) Ontological semantics, knowledge, ontologies, ontologically oriented methods of information analysis of texts. FIZMATLIT, Moscow
15. Zakharov VP, Mochalova AV, Mochalov VA (2016) Ontology Modification Using Ontological-Semantic Rules. ICACT Trans Adv Commun Technol (TACT) 5(5):902–906
16. Answers database. https://baza-otvetov.ru/
17. Mail.ru answers. https://otvet.mail.ru/
18. Zolotova GA (1988) Syntactic dictionary. The repertoire of the elementary units of Russian syntax (in Russian). Moscow, Russia: Nauka
19. Kaushinis TV, Kirillov AN, Korzhitsky NI, Krizhanovsky AA, Pilinovich AV, Sikhonina IA, Spirkova AM, Starkova VG, Stepkina TV, Tkach SS, Chirkova JV, Chuharev AL, Shorets DS, Yankevich DY, Yaryshkina EA (2015) A review of word-sense disambiguation methods and algorithms: introduction (in Russian). In: Proceedings of KarRC RAS. No. 10. Ser. mathematical modeling and information technologies, pp 69–98
20. Navigli R (2009) Word sense disambiguation: A survey. ACM Comput Surv (CSUR) 41(2, Article 10)

Seeding Programming

Vladimir Mochalov

Abstract The work is aimed at formalizing the implementation of the steps of the new method "seeding programming" focused on solving some optimization problems. Michelangelo told that there is a statue in every stone and all that is needed is to be able to remove all unnecessary and to take the statue to light. Based on Michelangelo's statement in the proposed method, we search for such a sequence of elements to remove from the original space ("stone"), which will lead to the formation of a set of remaining undeleted elements with the desired objective function. Initial elements of the search space either can be specified or they can be searched using special covering algorithms. To search for the sequence of elements to remove from the search space, we suggest to use search agents that form and use shared global memory.

Keywords Optimization method · Knowledge-based multi-agent system
Synthesis of solutions

1 Introduction

Michelangelo di Lodovico Buonarroti Simoni told that there is a statue in every stone and all that is needed is to be able to remove all unnecessary and to take the statue to light (Figs. 1 and 2). Statues differ from each other in appearance. Looking at a statue, we see the final result of the sculptor's work and speaking in technical language, we see the final objective function of the process of removing unnecessary elements ("small parts of the stone") from the original space ("stone"). Knowing the objective function, we can start to search for such extra elements from the initial space in order to obtain the desired result in the remainder.

V. Mochalov (✉)
Institute of Cosmophysical Research and Radio Wave Propagation FEB RAS,
Mirnaya str. 7, Kamchatka Region, Elizovskiy, Paratunka 684034, Russia
e-mail: sensorlife@mail.ru

R. Cárdenas et al. (eds.), *Proceedings of the 2nd International Conference on BioGeoSciences*, https://doi.org/10.1007/978-3-030-04233-2_5

Fig. 1 Michelangelo took the marble stone, removed all unnecessary, and got the statue of David

Fig. 2 Marble statues of the sculptor Bonazza located in the city of Peterhof

We draw the following analogies: “stone”—initial space from which we will remove the extra elements; “statue” is the best of the found solutions satisfying the given objective function; “sculptor” is a search agent that on the basis of knowledge and experience removes unnecessary elements from the initial space in order to obtain in the remainder a solution that satisfies the specified objective function. So, based on Michelangelo’s statement in the proposed method “Seeding programming,” we search for such a sequence of elements to remove from the original space (“stone”), which will lead to the formation of a set of remaining undeleted elements with the desired objective function. Initial elements of the search space can either be specified, or they can be searched using special covering algorithms. To search for the sequence of elements to remove from the search space, we suggest to use search agents.

2 Stages of Solving Problems Using Seeding Programming

Stage 1: Create objectives tree for solving optimization problem, define optimization parameters, define functions of estimating various optimization parameters, and form general target function and stopping criterion.

A concept of the “objectives tree” was introduced by Churchman and Ackoff in 1957. An objectives tree is a structure, constructed on the hierarchy principle (distributed into levels, ranged) assembly of project objectives, in which the following ones are emphasized: the general objective (“tree root”) and the subgoals of the first, second, and consequent levels subject to it (“tree branches”). In Fig. 3, a generalized objectives tree is shown. In leaf nodes of the tree, simple tasks are formed. Often the simple tasks are the requirements on achieving the specified thresholds of optimization parameters.

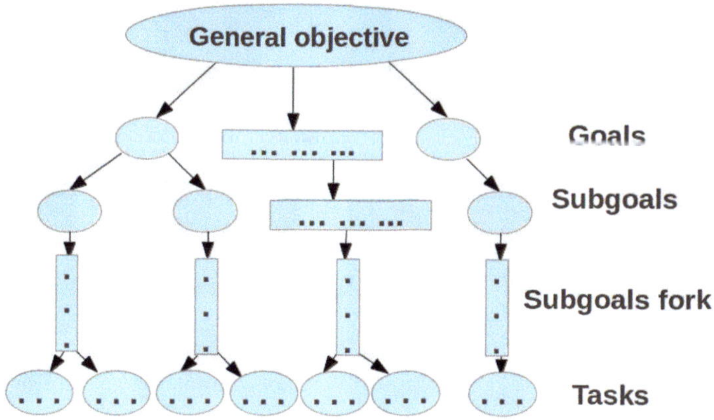

Fig. 3 Objectives tree

Stage 2: Select a suitable initial space ("stone"), removing elements from which we will look for the solution we need. It is necessary to mark that the smaller initial space contains less excess elements, so the less work required to remove them. At the same time, the initial space with a large number of elements potentially can be a greater number of better solutions compared to a space containing fewer elements. It is necessary to note the importance of suitable initial space selection. For example, if we choose the initial space that is too small, then we cannot find a solution that satisfies us, and vice versa, if we select too big initial space, then we will search for satisfying us solution for a very long time. We assume that elements of the initial space are either specified or their search is performed with special covering algorithms.

One of the interesting approaches to search the initial space is a knowledge-based multi-agent method for finding the sequence of adding elements in order to form a suitable initial space that contains the solution we need (ideally the initial space from which nothing needs to be removed). In this case, each agent adds elements to the start space on the basis of it rules of moving and general knowledge of the problem until it forms a set of elements which contains required solution after we can run seeding programming method to remove unnecessary elements (Fig. 4). For example, in paper [1] joint use of strategies for sequentially add elements to search space and sequentially remove elements from initial space are considered.

Stage 3: Create a shared global memory of agents (SGMA) for storing the agents' knowledge and experience about travelled routes (e.g., in [1] as SGMA is proposed to use shared global memory of the stored pheromone). Create an empty set of best

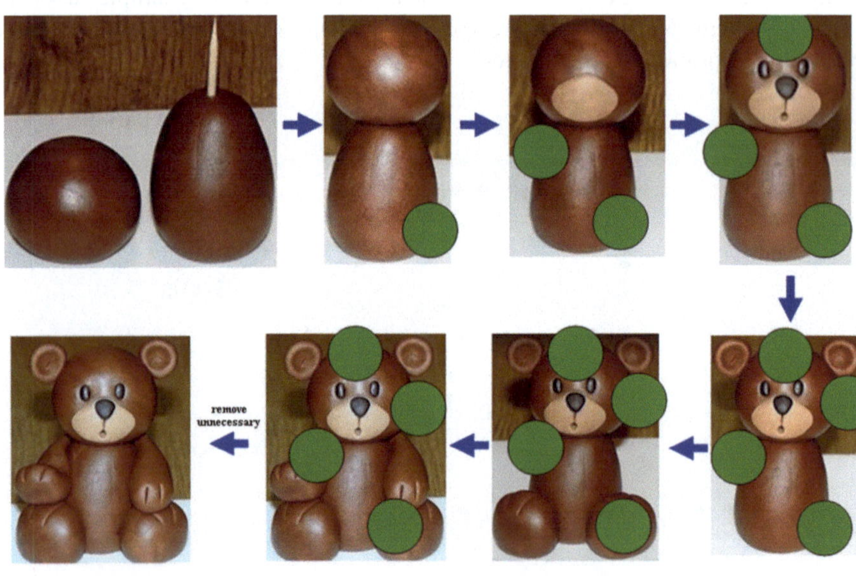

Fig. 4 Illustration of sequentially addition of elements to the start space

solutions Ω_{BEST}. Determine the maximum number of solutions b_K that will be stored in Ω_{BEST}.
Stage 4: Define agents that will be used to search unnecessary elements in initial space. Set for each agent parameters of its operation, objective function, and lower bound estimate K_{MIN} of the objective function.
Stage 5: Perform the following steps for each agent:

5.1. Using the movement rules of the agent and the SGMA, form the route M_T (array of elements) of agent moving on initial space elements,
5.2. Flip M_T array left to right,
5.3. In a loop for each element $T_O \in M_T$ perform these steps: (a) temporarily exclude T_O from M_T; (b) compute the confidence factor K_D of meeting the requirements of the objective function; (c) if in absence of T_O, the condition $K_D < K_{MIN}$ is satisfied then put T_O back to M_T into its place,
5.4. Using remaining elements in M_T, compute the confidence factor K_{DALL} of meeting the requirements of the overall objective function. If $K_{DALL} > 0$, then using some information about the remaining elements in M_T update SGMA.

Stage 6. If the stopping criterion is not met and if it is necessary, then update SGMA (e.g., in paper [1] shared global memory of the stored pheromone is updated using the following rule: $\Delta\tau_{ij}(t + 1) = (1 - p) * \tau_{ij}(t) + \Delta\tau_{ij}(t)$ [2]), reduce initial solution search space, and move to the stage 4. Otherwise return the best solution from Ω_{BEST}.

3 Seeding Programming Implementation for Synthesis of a Given Category Nodes Placement into Geospace Question–Answering Sensor Network Structure

New space technologies (nanosatellites, CubeSats, SmallSats, etc.), private space companies and the projects for launching thousands of small satellites to organize space networks with different purposes give principally new opportunities to monitor geospheres. We should note the increasing number of separate monitoring systems applying the data obtained from the geospace. The term "geospace" is understood as the region of space that goes from the solar photosphere to the atmosphere of Earth. It includes the solar photosphere, chromosphere and corona, the solar wind, Earth's magnetosheath, magnetosphere, thermosphere, ionosphere, and atmosphere.

In this paper on the basis of the personal results obtained before in the area of sensor networks construction [1, 3–6], semantic analysis, and question–answering systems [7–9], we introduce a new notion of "geospace question–answering sensor networks" (GQASN) that means a distributed network which monitors ambient environment parameters applying the data from geospace and allowing nodes to answer defined types of natural language questions as well.

In the work, we use a model of the GQASN structure (Fig. 5), where on the functional level the following types of GQASN nodes can be defined: (1) functional nodes (F-nodes) that collect information in some neighborhood of their location; (2) transit nodes (T-nodes) that manage routing and retransmit the information collected by F-nodes to the information collection centers (ICC) to be utilized further; (3) ICCs that manage the GQASN and process information collected by the GQASN. In general case, there can be multiple ICCs in the GQASN, and the information that has arrived into each of them is available to one or multiple users for making decisions and performing certain actions. It means that information received by F-nodes should be retransmitted, with a required degree of reliability, to several ICCs by means of transit nodes allocated within the given object in a certain way. We think that ICC is capable of performing F-node and T-node functions. F-node can perform the T-node functions and information between nodes can be transmitted both via the wire and wireless networks.

Designing of the GQASN requires the solution of many complicated problems referring to different areas of research; they are: projecting of network nodes (measurement stations, sensors, etc.); construction of different physical–mathematical models of monitoring processes of ambient environment parameters applying the geospace data; model selection for information collection from the GQASN; development of methods and algorithms for the GQASN structure synthesis; estimate of measurement error and limitations, estimate of spatial and other limitations of network node placement; ensuring of the defined functional and

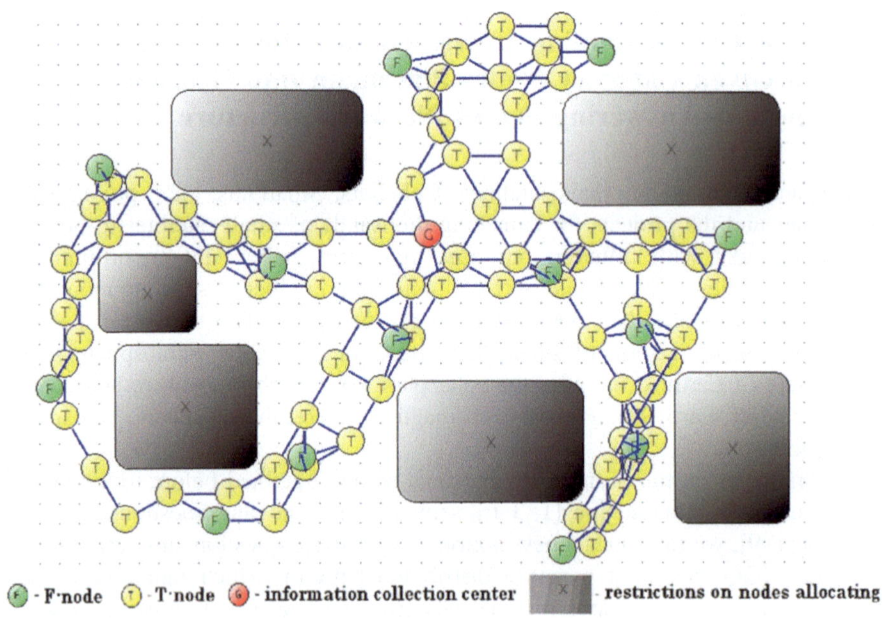

Fig. 5 Example of a distributed fault-tolerant GQASN structure

structural parameters of the synthesized GQASN; development of fitness functions of network nodes placement; development of algorithms for question–answering agents placement into the GQASN structure; development of self-organization algorithms for different GQASN levels (e.g., the function level performed by the nodes, routing level, level of tasks distribution between nodes and question–answering agents).

Two types of GQASN structures can be distinguished:

- distributed fixed network in which all the GQASN nodes are not moved after the initial placement;
- decentralized mobile self-organizing network compound of fixed nodes (distributed fixed network segment) and mobile units (mobile network segment) which can be moved in different directions and, as a consequence, form different network structures in dynamics by break and establish network connections with other nodes, removal and installation of new nodes into the network structure. Mobile units can be installed both on satellites and on mobile robots (drones, above-water and under-water vehicles). Some functions of the mobile units are: formation of a self-organizing GQASN structure; geographically distributed acquisition of data from the GQASN nodes; organization of the interaction of the GQASN mobile network segment with a fixed one. When applying the mobile robots, the following is possible: accurate nodes placement, distribution nodes over territory; moving, removing, reprogramming of nodes; charging and replacement of the GQASN nodes power sources; planning of cooperative behavior of mobile robots in the process of general aim solution and the GQASN nodes replacement based on the aims and current measurements of the whole GQASN.

Possible ways of problem statement for the GQASN structure synthesis

1. Synthesis of ICCs allocation. In this task, we know the spatial restrictions for allocating the ICCs. Also, we know the allocation of pre-installed ICCs. It is necessary to allocate ICCs in such way that the designed GQASN structure would have the "desired properties" assigned by a designer. During the synthesis of ICCs allocation, it is possible to optimize (by their removal or moving) some pre-installed ICCs noted by the designer.
2. Synthesis of F-nodes allocation. We know the spatial restrictions for allocating the F-nodes. Also, we know the allocation of ICCs and allocation of pre-installed F-nodes. It is necessary to allocate new F-nodes in such way that the designed GQASN structure would have the "desired properties" assigned by a designer. During the synthesis of F-nodes allocation, it is possible to optimize (by their removal or moving) some pre-installed F-nodes noted by the designer.
3. Synthesis of T-nodes allocation. We know the description of the GQASN allocation object, spatial restrictions for allocating the T-nodes. Also, we know the allocation of F-nodes, ICCs, and pre-installed T-nodes. It is necessary to allocate T-nodes in such way that the designed GQASN structure would have the "desired properties" assigned by a designer. During the synthesis of T-nodes

allocation, it is possible to optimize (by their removal or moving) some pre-installed T-nodes noted by the designer.
4. Complex sequential synthesis of ICCs, F-nodes, and T-nodes allocation. This statement suggests a sequential allocation of ICCs first (statement 1), then F-nodes (statement 2), and then T-nodes (statement 3).

It should be noted that the search space for synthesized solutions of concrete GQASN structure is very large and there are complex constraints in the objective function, and many of the solvable problems are NP-complete and to search for exact and approximate solutions of these problems, various algorithms of artificial intelligence [1, 10–13], linear and integer programming [14–16] are currently used and distributed calculations are performed.

Figure 6 illustrates a functional scheme of a given category nodes placement into the GQASN structure. This scheme can be used as the basis for synthesis of ICC, F-nodes, and T-nodes placement.

The question–answer agent (QA-agent) performs the function of generating the answer from natural language questions by collection, aggregation, and accumulation information from some F-nodes that are serviced by this QA-agent. After the accumulation of sufficient information from the group of F-nodes, the QA-agent generates an answer. The QA-agents can interact with each other to be able to answer the given types of questions under the established limitations. Physically, the QA-agent is a software/hardware add-on that can upgrade any type of GQASN nodes. The QA-agents can differ from each other by technical capabilities (due to various hardware and software add-ons) and functionality capabilities (the ability to answer different types of questions, performed functions, etc.).

In general, formulated in the natural language question Q enters to the input of one of the QA-agents that perform the functions of task coordinator for other QA-agents. After it the question Q enters to the input of the semantic analyzer module, which create ontological-semantic graph $G(Q)$. The graph $G(Q)$ enters to the input of the module for selection of QA-agents, which are best suited for generating the answer to question Q. The information about selected QA-agents is placed into a set FQA= {FQA_1, FQA_2, FQA_3, …, FQA_k}. The set FQA enters to the input of generating requests for QA-agents module. As a result, for each QA-agent $FQA_i \in FQA$, this module forms a request q_i. Each q_i request is transmitted to the QA-agent FQA_i, which first tries to find the necessary information in the local database and if it is not found FQA_i select serviced F-nodes from which the necessary information should be collected. The information of selected at this stage F-nodes is placed into the set $F = \{F_1, F_2, F_3, \ldots, F_n\}$. Further, the task formation module generates a task t_j for each F-node $F_j \in F$. Each F-node $F_j \in F$ that receives the task t_j executes it (using information stored on this F-node or receive it from the environment with the help of a sensor installed on the F-node) and sends the response r_j back to the QA-agent which generated the task t_j. On the basis of all the responses $r_1, r_2, r_3, \ldots, r_n$ obtained from F-nodes $F_1, F_2, F_3, \ldots, F_n$ QA-agent FQA_i generates a_i answer and sends it back to the task assignment

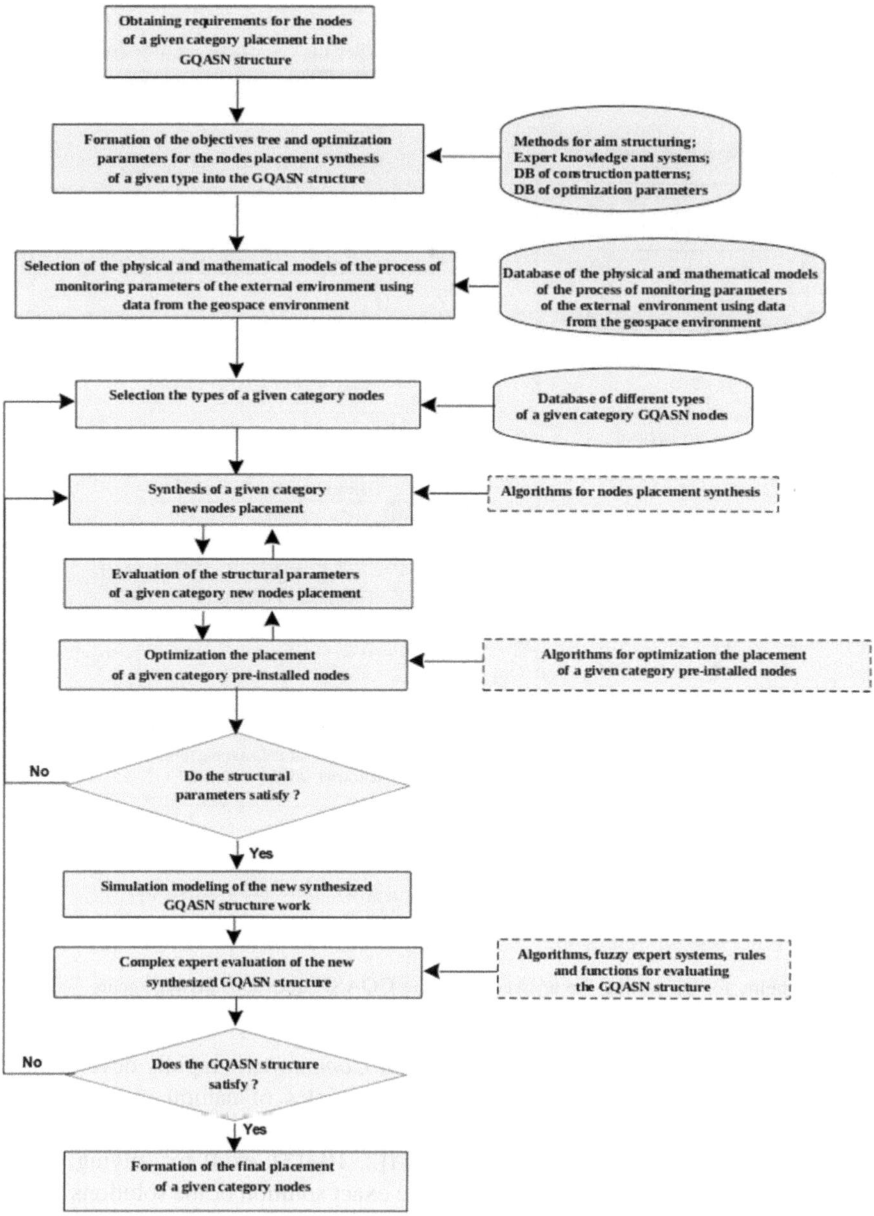

Fig. 6 Functional scheme of a given category nodes placement into GQASN structure

coordinator. Thus, the task assignment coordinator receives all answers from QA-agents and on the basis of them makes up a general answer *A*, which is transmitted to the user as an answer to the question *Q* (Fig. 7).

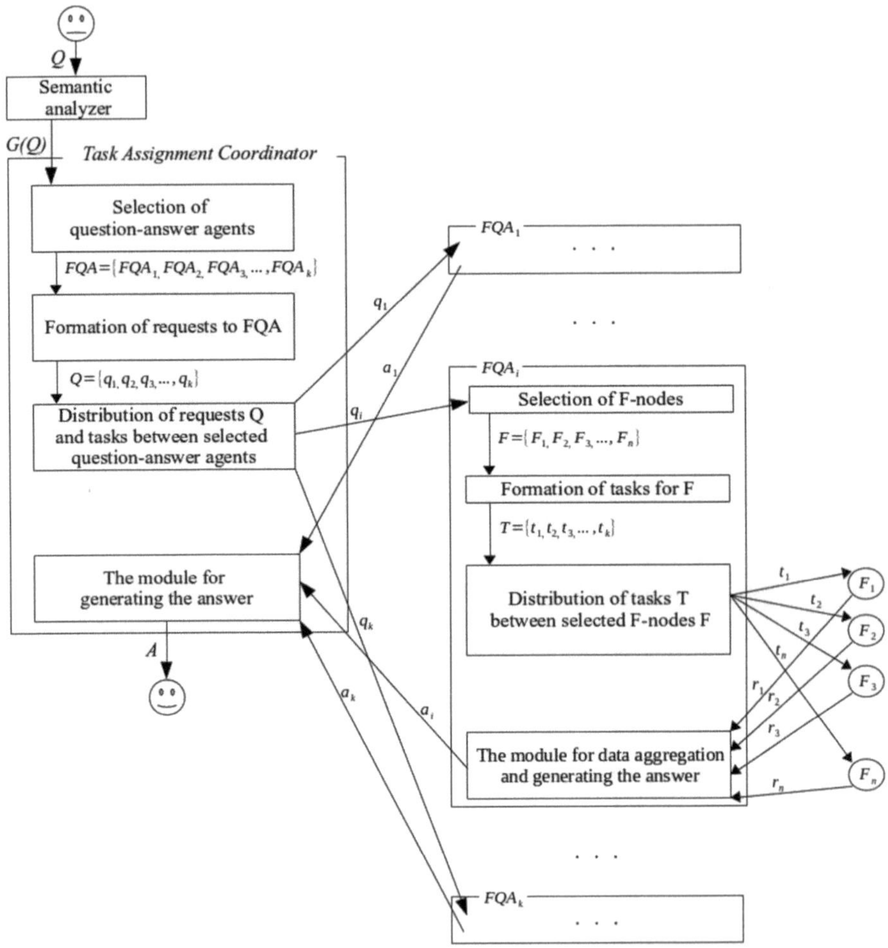

Fig. 7 Scheme for generating the answer using the GQASN question–answer agents

In recent years, the research area of *Natural Computing* is rapidly developing. It unites mathematical methods in which the principles of natural mechanisms of decision making are embedded [2]. Scientists have developed bio-inspired algorithms (BA) of modeling animals' behavior ([2, 10–13], etc.]) for solving various optimization problems that either do not have exact solution or the solutions' search space varies large and complex constraints of the objective function are presented, as well as NP-complete.

The described recommendations on applying BA and the proof in [17] that even the constrained variant of the problem of minimal coverage on plane is NP-complete allow us to conclude about the possibility to apply bio-inspired algorithms for the GSASN structure design.

Bio-inspired algorithms can be seen as multi-agent systems, each agent in which operates autonomously on very simple rules. The most frequently used bio-inspired agents (B-agents) include: ants, bees, termites, fireflies, birds, fish, bats, cats, and wolves.

The initial data of the algorithm for synthesis of a given category nodes $type_x$ placement are the following: allocation of pre-installed nodes of $type_x$ (set of nodes Ω_{pin}); allocation of installed nodes of type $\neq type_x$; description of the object that the GQASN needs to be located at (its dimensions, scheme, spatial requirements for nodes allocation and etc.); characteristics of ready-to-use nodes of $type_x$; adopted self-organization and routing algorithms; information collection model; functional requirements; optimization parameters; fuzzy expert systems, etc.

The algorithm below is based on the adaptation of the multi-agent bio-inspired algorithm for wireless sensor network design proposed in paper [1].

Step 1 Create objectives tree for solving optimization problem. Define a set M_{ALL} of all optimization parameters; the functions for calculating the parameters of M_{ALL}; a subset of optimization parameters $M_1 \in M_{ALL}$. Determine the membership functions of fuzzy sets that characterize the optimization parameters of the M_{ALL}; fuzzy expert system to derive the confidence factor to meet the functional requirements of the designer. Create an empty set of the best solutions Ω_{BEST}. Determine the maximum number of solutions b_K that will be stored in Ω_{BEST}.

Step 2 Create a set Ω_p of the possible placement points of type $type_x$ nodes (the set Ω_p can be formed with the help of: the algorithms for covering the object of placement with a mesh (based on an equilateral triangle or hexagon or square) or with circles with a given radius; covering algorithms in accordance with the choice and recommendations of the designer; other covering algorithms). Create an empty set Ω_T. Create node of type $type_x$ in each point of Ω_p and add this node to Ω_T set.

Step 3 Create a shared global memory of the stored pheromone (SGMSP) to share some "knowledge" between B-agents. The pheromone is stored on the edges of a fully connected undirected weighted graph (FCUWG), the nodes of which are the ones of $type_x$. To store the edges of the graph in computer memory, it is required to create a two-dimensional array *feromoneNetwork* with $N(N-1)/2$ memory cells of type *float*, where $N = |\Omega_T|$ is the number of nodes. All values of *feromoneNetwork* must be initialized as zeros.

Step 4 Define agents that will be used.

Step 5 Execute bio-inspired multi-agent algorithms.

5.1. Create a two-dimensional array *feromoneDif* to store changes in the pheromone using, for example, the following Java code:

```
float feromoneDif [ ] [ ] = new float[N-1] [ ];
for(int i = 0; i < N; i++)
```

```
feromoneDif [i] = new float[N - (i+1)];
```

All values of *feromoneDif* must be initialized as zeros (in the above code, the zero initialization is done automatically). Define the number of different bio-inspired agents *m,* the strategy for choosing the initial location of the agent, and other parameters needed for the agent to perform the work.

5.2. For each agent, perform the following steps (the code can be parallelized, i.e., to run in a separate thread for each agent):

5.2.1 Form, using the movement rules of the B-agent, SGMSP the route M_T (array) of agent moving on nodes Ω_T.

5.2.2. Create an empty extensible array of nodes M_{STR}, in which the nodes of the designed structure will be placed.

5.2.3. Select the design strategy:

(a) sequentially add nodes to the network -> Go to step 5.2.4;

(b) sequentially remove nodes from the network -> Add to M_{STR} all nodes from M_T array in the same sequence order. Go to step 5.2.11.

5.2.4. Create an empty set H_P, which will contain the caches of such internal parameters of the functions of computing estimates M_1, which will increase the speed of computing estimates M_1 for the next iteration. Create a variable i to store the index of the current node from the M_T array and initialize its value to 0 ($i = 0$). Set the node T_C ($T_C = M_T[0]$) as the current one.

5.2.5. Add to M_{STR} the node T_C. Form the network structure S_S of nodes M_{STR}.

5.2.6. Calculate using the caches H_P the values estimations of the optimization parameters of the set M_1 having structure S_S. Clear H_P. Save the caches of the internal parameters of the functions of computing estimates M_1 of the current iteration to the set H_P.

5.2.7. Using a fuzzy expert system, calculate the reliability coefficient K_{D1} of meeting the requirements of the designer for parameters of a set M_1 of structure S_S.

5.2.8. If $K_{D1} > p_1$, where p_1 is a set threshold, go to step 5.2.11.

5.2.9. If $i < |M_T|$, put $i = i + 1$ and accept the next node $T_C = M_T[i]$ as the current one. Repeat steps 5.2.5.–5.2.9. while i does not become equal to $|M_T|$.

5.2.10. Exit with notification of the failure from the function of agent design of the network structure.

5.2.11. Steps of eliminating optimization:

5.2.11.1. Select the strategy of eliminating optimization:

(a) step-by-step optimization with consideration of optimization parameters M_1. The fuzzy expert estimation of the structural parameters M_1 is used;

(b) step-by-step optimization with consideration of all optimization options M_{ALL}. The unit of simulation modeling and complex assessment of the network is used;

5.2.11.2. Revert the M_{STR} array,

5.2.11.3. In a loop, temporarily exclude each node $T_O \in M_{STR}$ from M_{STR}, then compute the confidence factor K_D of meeting the requirements of the eliminating optimization strategy parameters. If in absence of node T_O evaluation of network structure stops meeting the designer requirements, put T_O back to M_{STR} into its place.

5.2.11.4 Revert the M_{STR} array,

5.2.12. Perform simulation modeling of the network. The results of the modeling and structural-parametric estimates of the various parameters are the input to the complex expert system for evaluation of network structure. Calculate with the latter the confidence factor K_{DALL} of meeting all the designer requirements.

5.2.13. If $K_{DALL} > 0$ then, in accordance with one of the following strategies, increase the pheromone amount in the array *feromoneDif*:

(a) consequent update—increase the amount of pheromone on the edges of the agent sequential traveling on nodes of M_{STR} by the value equal to $\Delta\tau_{ij,k}(t) = Q_{agent} * K_{DALL}$, where Q_{agent} is the amount of pheromone secreted by the agent on one edge;

(b) full-mesh update—increase the amount of pheromone on all edges of the fully connected graph constructed on the basis of nodes of M_{STR} by the value equal to $\Delta\tau_{ij,k}(t) = Q_{agent} * K_{DALL}$.

5.2.14. If K_{DALL} is greater than the estimate of the worst solution from Ω_{BEST}, or ($|\Omega_{BEST}| < b_K$ and $K_{DALL} > 0$), then add into Ω_{BEST} the current solution. By the solution, we mean the couple (M_{STR}, K_{DALL}). If $|\Omega_{BEST}| \geq b_K$, then leave in Ω_{BEST} only b_K best solutions.

5.3. After all agents have performed step 5.2, update SGMSP (*feromoneNetwork* array) in accordance with the following well-known rule [2]: $\Delta\tau_{ij}(t + 1) = (1 - p) * \tau_{ij}(t) + \Delta\tau_{ij}(t)$, where $\Delta\tau_{ij}(t)$ is the amount of pheromone on edge (i, j) in the array of pheromone changes *feromoneDif*, and $p \in [0, 1]$ is the coefficient of pheromone evaporation. To enhance the intermediate best solutions, the amount of pheromone on the edges of the routes of the best solutions Ω_{BEST} should be increased (an example is using "elite" ants).

5.4 If the stopping criterion is not met, go to step 5.1.

Step 6 If it is necessary to continue the search, then create new set Ω_T and add to it type type$_x$ nodes located in points of possible nodes placement (e.g., to cover the object of nodes placement with a mesh more densely in comparison with the previous coverage) and move to Step 3. Otherwise return the best solution from Ω_{BEST}.

4 Conclusion

Seeding programming is a new method for solving various optimization problems that either do not have exact solution or the solution search space is very large and complex constraints of the objective function are present, as well as NP-complete.

One possible implementation of the algorithm that implements the seeding programming method is given in this paper for synthesis of a given category nodes placement into the GQASN structure.

References

1. Mochalov VA (2015) Multi-agent bio-inspired algorithms for wireless sensor network design. In: Proceedings on IEEE 17th international conference on advanced communication technology, ICACT 2015, Phoenix Park, Korea, pp 34–42
2. Bonavear F, Dorigo M, Theraulaz G (1999) Swarm intelligence: from natural to artificial systems. Oxford University Press, New York, 320 p
3. Mochalov VA (2015) Synthesis of the wireless sensor network structure in the presence of physical attacks. Lect Notes Comput Sci 9247:11–22
4. Mochalov VA, Pschenichnikov AP (2015) Functional scheme of the flying sensor networks architecture design. ICACT Trans Adv Commun Technol (TACT) 4(4):659–663. https://doi.org/10.1109/ICACT.2016.7423591
5. Mochalov VA (2016) Certificate of registration of computer software "Program for synthesis of monitoring networks by bio-inspired algorithms" No. 2016612039
6. Mochalov VA, Mochalova AV (2017) Algorithms for changing the structure of geospace self-organizing question-answering sensor networks. In: VIII international conference "solar-terrestrial relations and physics of earthquake precursors", p 11. https://doi.org/10.1051/e3sconf/20172002009
7. Mochalova AV, Mochalov VA (2017) Mathematical model of an ontological-semantic analyzer using basic ontological-semantic patterns. In: Lecture notes in artificial intelligence, proceedings of 15th Mexican international conference on artificial intelligence, pp 53–66
8. Mochalova AV, Zacharov VP, Mochalov VA (2017) Ontology modification using ontological-semantic rules. In: 19th international conference on advanced communications technology (ICACT)—opening new era of smart society, pp 902–906
9. Kuznetsov VA, Mochalov VA, Mochalova AV (2016) International conference on advanced communication technology, ICACT, pp 651–658
10. Brabazon A, O'Neill M, McGarraghy S (2015) Natural computing algorithms. Springer, 554 p
11. Mandal JK, Mukhopadhyay S, Pal T (2016) Handbook of research on natural computing for optimization problems (Vols 2), Igi-Global, 1015 p
12. Fister I Jr, Xin-She Y, Fister I, Brest J, Fister D (2013) A brief review of nature-inspired algorithms for optimization. ELEKTROTEHNIˇSKI VESTNIK 80(3):1–7
13. Binitha S, Sathya S (2012) A survey of bio inspired optimization algorithms. Int J Soft Comput Eng (IJSCE) 2(2):137–151
14. Gounaris CE, Rajendran K, Kevrekidis IG, Floudas CA (2015) Designing networks: a mixed integer linear optimization approach, 56 p. https://arxiv.org/pdf/1502.00362.pdf
15. Taccari L (2015) Mixed-integer programming models and methods for bilevel fair network optimization and energy cogeneration planning. Ph.D. dissertation, 209 p

16. Fraccaroli E, Quaglia D Toolchain for optimal network synthesis. http://www.di.univr.it/documenti/OccorrenzaIns/matdid/matdid014072.pdf
17. Fowler RJ (1981) Optimal packing and covering in the plane are NP-complete. Inf Process Lett 12(3):133–137

Schwarzschild Metric Disturbed by the Accelerated Expansion of the Universe

Adrian Linares-Rodríguez, Carlos R. Fadragas and Ailier Rivero-Acosta

Abstract The purpose of this work was to analyze the static spherically symmetric solution of the Einstein field equations for the case of a massive static spherical object surrounded with quintessential matter. A new approach was considered where the energy–momentum tensor was maintained null, and the effect of the quintessence (accelerated expansion of the Universe) was directly introduced by modifying the metric tensor. We deal the case of quintessence with the state parameter $w = -2/3$. The results obtained are in correspondence with other approaches present in the bibliography; for instance, the studies carried out about the same problematic in Kiselev (Quintessence and black hole, 2002 [1]), where they got a new static spherically symmetric solution through the energy–momentum tensor for a black hole surrounded with quintessential matter.

Keywords Schwarzschild solution · Kiselev black hole · Accelerated expansion Quintessence · Dark energy

1 Introduction

In the Universe, extraordinary events occur with great difficulty understood by man. Sometimes, these events can be very powerful phenomena, and the Universe can become an inhospitable and unpredictable scenario. Any form of life on our planet or any other planet outside the solar system is threatened constantly by these cosmic phenomena within this colossal scenario.

The main characteristics of each phenomenon are always associated with the scale in which it takes place. That is to say, the Universe shows different aspects and magnitudes of the physical parameters in different scales. In this way, various material structures can be observed, and many of these structures can find a plausible explanation based on scientific knowledge. However, other structures still

A. Linares-Rodríguez (✉) · C. R. Fadragas · A. Rivero-Acosta
Universidad Central “Marta Abreu” de Las Villas, Villa Clara, Cuba
e-mail: adlinares@uclv.cu

R. Cárdenas et al. (eds.), *Proceedings of the 2nd International Conference on BioGeoSciences*, https://doi.org/10.1007/978-3-030-04233-2_6

hope to have a true explanation. We can observe, for example, the scales for the different physical magnitudes associated with our terrestrial life conditions. The habitability on our planet Earth and the existence of the various life forms in it show a significant peculiarity of the Universe. The conditions for life are formed by the Universe, and the Universe itself can take them away from us. In fact, everything seems to indicate that the conditions of habitability of our planet present some critical variations with respect to the favorable average values for the existence of the different species that we know. Our habitability conditions are even very susceptible to the secondary effects of the technological development itself.

On the other hand, it is an undeniable fact that the technological development that has reached human society has allowed to observe with much greater precision at different scales. Fundamentally the spatial, temporal magnitudes, and although a little less, also the energetic ones. The astronomical observations [2–7] show that the Universe expands rapidly. This accelerated expansion is another of the bewildering events of the cosmos. A Universe where only ordinary matter is supposed to exist should be contracting, because gravitational attraction would predominate. However, this reality has made theorists think about the existence of an extraordinary component known as dark energy (DE). This is a form of matter or energy that would be present in all space, producing a pressure that tends to accelerate the expansion, resulting in a repulsive gravitational force.

Consider the existence of dark energy is the path most used by researchers to explain the accelerated expansion observed. We must not confuse the dark energy with the dark matter. In the Standard Model of Cosmology, the dark energy contributes almost three-quarters of the total mass–energy of the Universe and the dark matter only about a quarter. The rest is baryonic matter (approximately 5%).

We know that the theory of gravitation shows that the Sun is the main body that modifies the spatial geometry of the solar system and forces the planets to move in well-defined orbits. On the other hand, relatively recent consensus was established by the scientific community that our galaxy must surely have a very massive structure at its center, known as a black hole. Then, on a galactic scale, we can consider that we largely inhabit a region over space-time defined by the central black hole metric. We all know that a black hole is an amazingly massive body, with an incredible gravitational force capable of dragging matter toward it and continue to increase its power infinitely. Therefore, it is undeniable that the black hole could influence in some way on life on our planet in the future.

Then, considering the ideas mentioned above, it is not inappropriate to study the metric of the central black hole together with the effect of accelerated expansion, with the purpose of getting closer to the reality that surrounds us.

However, we must not forget that, being members of the Universe, we can be affected in many ways both "from the outside" and "from within." In addition, our planetary conditions of habitability are in general quite sensitive to changes in the variables that describe such conditions. We should not abandon any line of study, since what we know about the Universe is just the beginning of everything we should know, if man will ever try to leave the cradle in which civilization took place in the future. In fact, covering all scales (both "outside" and "inside") is a necessity to have greater possibilities of preserving the human species. The Universe is evidently the most complex dynamic system that exists, so that it is increasingly necessary to take a step forward, to try not only to understand nature, but also to be able to act alongside it and try to do our will.

Let us see then as a step forward that the omnipresent existence of dark energy can have some effect on the space-time metric of a black hole. We will therefore introduce a spatial scale factor dependent on time, in the equations that describe the black hole metric. The scale factor $a(t)$ will be introduced in a similar way as in the Friedmann–Lemaitre–Robertson–Walker metric.

We will begin by citing the first exact solution of the Einstein equations developed by Karl Schwarzschild a few months after Albert Einstein established his General Theory of Relativity [8]. We will also introduce in the aforementioned way the presence of a field of quintessence surrounding the static black hole.

This field would be very light, and this implies a state of negative pressure. The origin of the negative pressure could be twofold. The first one is the cosmologic constant, and the second one is the quintessential with the state equation given by the relation between the pressure p_q and the density of energy ρ_q, being $p_q = \omega_q \rho_q$ in the interval from $-1 < \omega_q < -1/3$, depending on the cause of acceleration [1].

2 Exact Solutions in Static Coordinates

In the research [1], they give a general solution of the static spherically symmetric Einstein's equations with quintessential that satisfy the conditions of additively and linearity, which allow treating the problems with the charged or non-charged black hole in the flat space or De Sitter space [9].

They found a new form of exact solution static spherically symmetric Einstein's equations, which describe a black hole surrounded by quintessential through energy–moment tensor.

$$g_{tt} = 1 - \frac{r_s}{r} - \frac{C}{r^{3\omega_q+1}} \tag{1}$$

The metric is given by:

$$ds^2 = \left[1 - \frac{r_s}{r} - \frac{C}{r^{3\omega_q+1}}\right] dt^2 - \left[1 - \frac{r_s}{r} - \frac{C}{r^{3\omega_q+1}}\right]^{-1} dr^2 - r^2\left(d\theta^2 - \sin^2\theta d\phi^2\right), \tag{2}$$

where $r_s = 2Gm/c^2$, considering the light's velocity $c \neq 1$, m the mass of the black hole, and C the normalization constant.

3 Quintessential with State Parameter $\omega = -2/3$

In the earlier solution, we can analyze the important limit cases in the asymptotically flat space and De Sitter. For example, the case of a charged black hole surrounded by the static spherically symmetric electric field corresponds to the case with the relativistic matter state parameter $\omega = 1/3$. In this case, the general solution [1] gives the Reissner–Nordström metric or the Schwarzschild metric, when the charge $Q = 0$, if we consider:

$$g_{tt} = -\frac{1}{g_{rr}},$$

we obtain:

$$g_{tt} = 1 - \frac{r_s}{r} + \frac{Q^2}{r^2},$$

and for $Q = 0$:

$$g_{tt} = 1 - \frac{r_s}{r},$$

There also exists the important case with $\omega_q = -2/3$, studied in detail in [1], where they obtained:

$$g_{tt} = 1 - \frac{r_s}{r} - \frac{C}{r^{3\left(-\frac{2}{3}\right)+1}} = \left[1 - \frac{r_s}{r} - \frac{r}{C}\right],$$

And in the free quintessential case:

$$g_{tt} = -\frac{1}{g_{rr}} = 1 - \frac{C}{r^{3\left(-\frac{2}{3}\right)+1}} = \left[1 - \frac{r}{C}\right].$$

4 Modified Schwarzschild Metric

Now let us work from the geometric part of the Einstein equation; that is, we start from a general proposal and taking into account that the DE is present in the model, we should consider the scale factor $a(t)$ related to the accelerated expansion in the spatial terms, in such a way that

$$ds^2 = e^{2A(r)}c^2dt^2 - a(t)^2e^{2B(r)}dr^2 - a(t)^2r^2\left(d\theta^2 + \sin^2\theta d\phi^2\right)$$

with $A(r)$ and $B(r)$ two functions that remain to be determined.

Table 1 Christoffel symbols

$\Gamma^{r}_{tt} = e^{2(A-B)}\frac{c^2}{a^2}A'$	$\Gamma^{\theta}_{r\theta} = r^{-1}$	$\Gamma^{\phi}_{\theta\phi} = \cot\theta$
$\Gamma^{t}_{tr} = A'$	$\Gamma^{\phi}_{r\phi} = r^{-1}$	$\Gamma^{r}_{\phi\phi} = -r\sin^2\theta \mathrm{e}^{-2B}$
$\Gamma^{r}_{rr} = B'$	$\Gamma^{r}_{\theta\theta} = -r\mathrm{e}^{-2B}$	$\Gamma^{\theta}_{\phi\phi} = -\sin\theta\cos\theta$
$\Gamma^{\theta}_{t\theta} = \frac{\dot{a}}{a}$	$\Gamma^{r}_{tr} = \frac{\dot{a}}{a}$	$\Gamma^{t}_{rr} = \mathrm{e}^{2(B-A)}\frac{a\dot{a}}{c^2}$
$\Gamma^{t}_{\phi\phi} = \mathrm{e}^{-2A}r^2\sin^2\theta\frac{a\dot{a}}{c^2}$	$\Gamma^{t}_{\theta\theta} = \mathrm{e}^{-2A}r^2\frac{a\dot{a}}{c^2}$	$\Gamma^{\phi}_{t\phi} = \frac{\dot{a}}{a}$

We substitute the previous proposal in the vacuum equations $R_{\mu v} = 0$ to determine A and B through the geometrical part of the Einstein equations (Table 1).

So the nonzero components of the Ricci tensor are

$$
\begin{aligned}
R_{\mathrm{tt}} &= -\frac{c^2}{a^2}A''\mathrm{e}^{2A-2B} - \frac{c^2}{a^2}A'^2A'\mathrm{e}^{2A-2B} + \frac{3\ddot{a}}{a} + \frac{c^2}{a^2}A'B'\mathrm{e}^{2A-2B} - \frac{2}{r}\frac{c^2}{a^2}A'\mathrm{e}^{2A-2B},\\
R_{\mathrm{rr}} &= A'' + A'^2 - A'B' - \frac{2}{r}B' - \frac{a^2}{c^2}\left(\frac{\ddot{a}}{a} + \frac{2\dot{a}^2}{a^2}\right)\mathrm{e}^{2B-2A},\\
R_{\theta\theta} &= -\frac{r^2}{c^2}\ddot{a}a\mathrm{e}^{-2A} - rB'\mathrm{e}^{-2B} + rA'\mathrm{e}^{-2B} - 2r^2\frac{\dot{a}^2}{c^2}\mathrm{e}^{-2A} + \mathrm{e}^{-2B} - 1,\\
R_{\phi\phi} &= \sin^2\theta R_{\theta\theta}
\end{aligned}
$$

Taking R_{tt} and R_{rr} from the above set and making

$$
R_{\mathrm{tt}}\frac{a^2}{c^2}\mathrm{e}^{2B-2A} + R_{\mathrm{rr}} = 2\left(\frac{\ddot{a}}{a} - \frac{\dot{a}^2}{a^2}\right)\frac{a^2}{c^2}\mathrm{e}^{2B-2A} - 2r^{-1}(A' - B')
$$

and then considering the relations

$$
H = \frac{\dot{a}}{a}, \quad \dot{H} = \frac{\ddot{a}}{a} - \frac{\dot{a}^2}{a^2} = \frac{\ddot{a}}{a} - H^2, \quad \dot{H} + H^2 = \frac{\ddot{a}}{a}, \tag{3}
$$

we get

$$
R_{\mathrm{tt}}\mathrm{e}^{2B-2A}\frac{a^2}{c^2} + R_{\mathrm{rr}} = 2\dot{H}\frac{a^2}{c^2}\mathrm{e}^{2B-2A} - 2r^{-1}(A' + B') \equiv 0
$$

If we divide by e^{2B-2A} and regroup, we obtain

$$
2\dot{H}\frac{a^2}{c^2} = 2r^{-1}(A' + B')\mathrm{e}^{2A-2B} = -\lambda = \Gamma, \tag{4}
$$

where

$$\Gamma = \text{constant} = 2\dot{H}\frac{a^2}{c^2},$$

therefore

$$-2r^{-1}(A' + B') = \lambda e^{2B-2A},$$

and if we consider the case without accelerated expansion and therefore with zero scale factor, $\lambda \equiv 0$, we obtain then

$$-2r^{-1}(A' + B') \equiv 0,$$

which leads to the ordinary solution of the Schwarzschild metric.

Considering the relations (3) and replacing in R_{rr}, we have

$$R_{\mathrm{rr}} = A'' + A'^2 - A'B' - 2r^{-1}B' - \frac{a^2}{c^2}\left(\dot{H} + H^2 + 2H^2\right)e^{2B-2A},$$

or better

$$R_{\mathrm{rr}} = A'' + A'^2 - A'B' - 2r^{-1}B' - \left(\frac{\Gamma}{2} + 3H^2\frac{a^2}{c^2}\right)e^{2B-2A},$$

and making

$$\Delta = \left(\frac{\Gamma}{2} + 3H^2\frac{a^2}{c^2}\right),$$

we obtain

$$\left(A'' + A'^2 - A'B' - 2r^{-1}B'\right)e^{2A-2B} = \Delta = \frac{\Gamma}{2} + 3H^2\frac{a^2}{c^2}.$$

Taking B' from (4) and replacing in the above equation, we obtain

$$\left(A'' + 2A'^2 + \frac{2}{r}A^2\right)e^{2A-2B} - \Gamma\left(\frac{rA'}{2} + 1\right) = \Delta \tag{5}$$

Using the same previous method on $R_{\theta\theta}$ from group (7), we obtain

$$\left(\frac{2A'}{r} + \frac{1}{r^2}\right)e^{2A-2B} - \frac{e^{2A}}{r^2} = \Delta + \frac{\Gamma}{2}. \tag{6}$$

Equations (5) and (6) confirm the system equations of model considered, and $A(r)$ and $B(r)$ are functions unknown, which must be found.

We made an analytical construction of the expression (6) starting from (1), considering

$$e^{2A} = \left(1 - \frac{r_s}{r} + \delta\right) = e^{-2B}, \tag{7}$$

So,

$$e^{2A}e^{-2B} = \left(1 - \frac{r_s}{r} + \delta\right)^2,$$

With $\delta = \delta(r)$, this being the new unknown function of the r

$$\begin{aligned}
&e^{2A}2A' = \left(0 + \frac{r_s}{r} + \delta\right),\\
&A' = \frac{1}{2}\left(\frac{r_s}{r^2} + \delta'\right)\left(1 - \frac{r_s}{r} + \delta\right)^{-1},\\
&\frac{2A'}{r} = \frac{1}{r}\left(\frac{r_s}{r^2} + \delta'\right)\left(1 - \frac{r_s}{r} + \delta\right)^{-1},\\
&\frac{2A'}{r} + \frac{1}{r^2} = \frac{1}{r}\left(\frac{r_s}{r^2} + \delta'\right)\left(1 - \frac{r_s}{r} + \delta\right)^{-1} + \frac{1}{r^2},\\
&\left(\frac{2A'}{r} + \frac{1}{r^2}\right)e^{2A-2B} = \frac{1}{r}\left(\frac{r_s}{r^2} + \delta'\right)\left(1 - \frac{r_s}{r} + \delta\right) + \frac{1}{r^2}\left(1 - \frac{r_s}{r} + \delta\right)^2,\\
&\left(\frac{2A'}{r} + \frac{1}{r^2}\right)e^{2A-2B} - \frac{e^{2A}}{r^2} = \left[\frac{1}{r}\left(\frac{r_s}{r^2} + \delta'\right) - \frac{1}{r^2}\right]\left(1 - \frac{r_s}{r} + \delta\right) + \frac{1}{r^2}\left(1 - \frac{r_s}{r} + \delta\right)^2,
\end{aligned} \tag{8}$$

where in the previous expression the left member of Eq. (6) has been built, so the resulting terms in the right member of (8) are also equal to the right member of the original Eq. (6), leaving

$$\left[\frac{1}{r}\left(\frac{r_s}{r^2} + \delta'\right) - \frac{1}{r^2}\right]\left(1 - \frac{r_s}{r} + \delta\right) + \frac{1}{r^2}\left(1 - \frac{r_s}{r} + \delta\right)^2 = \Delta + \frac{\Gamma}{2},$$

It follows

$$\begin{aligned}
&\left(\frac{r_s}{r^2} + \delta'\right) + \frac{1}{r}\left(1 - \frac{r_s}{r} + \delta\right) - \frac{1}{r} = \frac{r(\Delta + \frac{\Gamma}{2})}{1 - \frac{r_s}{r} + \delta},\\
&\delta' = \frac{r(\Delta + \frac{\Gamma}{2})}{1 - \frac{r_s}{r} + \delta} + \frac{1}{r} - \frac{1}{r}\left(1 - \frac{r_s}{r} + \delta\right) - \frac{r_s}{r^2},\\
&\delta' = \frac{r(\Delta + \frac{\Gamma}{2})}{1 - \frac{r_s}{r} + \delta} - \frac{\delta}{r},\\
&\delta' = \frac{r(\Delta + \frac{\Gamma}{2}) - \delta(1 - \frac{r_s}{r} + \delta)}{(1 - \frac{r_s}{r} + \delta)r},
\end{aligned}$$

where considering $K = \Delta + \Gamma/2$ and separating rK, we obtain

$$\left(\delta' + \frac{\delta}{r}\right)\left(1 - \frac{r_s}{r} + \delta\right) = rK, \tag{9}$$

From Eq. (9), we will determine $\delta(r)$, which according to (1) must have the form

$$\delta(r) = \frac{C}{r^{\alpha}}, \tag{10}$$

where $\alpha = 3\omega q + 1$

Taking (9) and multiplying by r, we get

$$(r\delta' + \delta)\left(1 - \frac{r_s}{r} + \delta\right) = r^2 K,$$

where we can see that $(r\delta' + \delta)$ is the derivative of a product, so

$$(r\delta)'\left(1 - \frac{r_s}{r} + \delta\right) = r^2 K,$$

or better

$$(r\delta)' = \frac{r^2 K}{\left(1 - \frac{r_s}{r} + \delta\right)} \tag{11}$$

From (11), limit cases of r of physical interest can be analyzed. These cases allow approximations that facilitate the integration of the expression.

For example, we have the notable case of $r \gg r_s$ very far from the spherical body, for which

$$(r\delta)' \approx \frac{r^2 K}{1 - \delta}.$$

As r_s/r would take a negligible value, then if you consider that C/r^{α} grows, for a $\alpha < 0$ and considering that r is very large, $\delta \gg 1$, so you can write

$$(r\delta)' \approx \frac{r^2 K}{\delta},$$

which we multiply by r and reorganize as,

$$r\delta \mathrm{d}(r\delta) \approx r^3 \mathrm{d}r K,$$

and integrating,

$$\delta \approx \pm\sqrt{K\frac{r^2}{2} + \frac{2K_1}{r^2}}.$$

The above expression being the sought forms of δ. For very large r, the term $2K_1/r^2 \approx 0$, remaining

$$\delta \approx \pm\sqrt{\frac{K}{2}}r, \tag{12}$$

where comparing with (10), it turns out that $\alpha = -1$ and $C = \pm\sqrt{K/2}$, which means that if $\alpha = (3\omega + 1)$, then $\omega = -2/3$, this value of the parameter of state ω with the one reported in [1].

Then, taking into account (7) and (11), it remains that

$$e^{2A} \approx \left(1 - \frac{r_s}{r} \pm \sqrt{\frac{K}{2}}r\right),$$

where, without forgetting that $r \gg r_s$, $r_s/r \approx 0$, so that

$$e^{2A} \approx \left(1 \pm \sqrt{\frac{K}{2}}r\right).$$

5 Conclusions

We have a new approach to present the first exact solution of Einstein equations for the static spherically symmetric quintessential state surrounding a black hole. The solution was obtained taking into account the presence of the quintessence around the black hole, by introducing a time-dependent scale factor, similar to how it was used in the Friedmann–Lemaitre–Robertson–Walk metric. We assume null the energy–momentum tensor contribution to Einstein equations and evaluate correspondence of results here with those obtained by Kiselev. In particular, for quintessence field with the state parameter $\omega = -2/3$, there exist complete correspondence with that results obtained by Kiselev.

With the approaching considered here, we have avoided the use of Lagrangian formalism for a scalar field with a specific potential and a kinetic term, and just we have introduced an effect given by the expanding of the Universe. Other conditions, for example, the no-hair theorem remaining valid, were maintained. We assume that exact solutions for a charged black hole or for a rotating black hole may be considered by the same way, and they can be modified introducing the scale factor depending on time in the spatial terms of the corresponding metric. It is important to observe that with this approaching used here, the particular type of dark energy is not an interesting issue.

References

1. Kiselev VV (2002) Quintessence and black hole. arXiv:gr-qc/0210040v3
2. Perlmutter S et al (1999) Supernova cosmology project collaboration. Astrophys J, 517, 565 [arXiv: astro-ph/9812133]; Riess AG et al (1998) Supernova search team collaboration. Observational evidence from supernovae for an accelerating universe and a cosmological constant. Astron J 116: 1009–1038 astro-ph/9805201v1
3. Schmidt BP et al, The high-Z supernova search: measuring cosmic deceleration and global curvature of the universe using type IA supernovae. arXiv: astro-ph/9805200v1
4. Kowalski M et al (2008) Supernova cosmology project collaboration. Astrophys J 686(2):749. arXiv: 0804.4142v1 [astro-ph]
5. Allen S et al (2008) Improved constraints on dark energy from Chandra X-ray observations of the largest relaxed galaxy clusters. Mon Not Roy Astron Soc 383(3):879–896. arXiv: 0706.0033v3 [astro-ph]
6. Abazajian KN et al, The seventh data release of the sloan digital sky survey. arXiv: 0812.0649v2[astro-ph]
7. Ade PAR et al, (Planck collaboration): planck 2013 results. XXII. Constraints on inflation. arXiv: 1303.5082v3 [astro-ph.CO]
8. Janssen B (2013) Teoría de la Relatividad General
9. Kofane TC, Bouetou TB, Saleh M (2012) Thermodynamics and phase transition of the reissner-nordström black hole surrounded by quintessence. Gen Relat Gravit

Potential Changes on Anammox Activity After Chicxulub Asteroid Impact

Noel Perez, Osmel Martin and Rolando Cárdenas

Abstract The activity of anammox bacteria (AA) can be estimated based on the production of N_2. Known the mechanism of the catabolic reaction, three differential equations are established according to the most important substrates (NO_3^- and NH_4^+) and the pH. The solution of this system might provide a rough estimate of the activity of the anammox, since neither inhibitory effects nor efficiency of the microorganisms is considered. Anammox can be inhibited by ocean acidification, generated by the deposition of H_2SO_4 and HNO_3 during asteroid impact events such as Chicxulub. The magnitude of the inhibition depends on the amount and speed with which the H_2SO_4 is added fundamentally.

Keywords Anammox activity · Asteroid impact · Ocean acidification

1 Introduction

The process of anoxic oxidation of ammonium (anammox) is one of the main processes of biological conversion of nitrogen in nature. It is characterized by the reaction of nitrite with ammonium in anoxic condition to form dinitrogen gas. The study of the anammox process began at the end of the twentieth century. The existence of this process in itself was a great surprise because it was discovered more than one hundred years after the other basic players in the nitrogen cycle, such as nitrogen fixation, denitrification, and nitrification, had been identified. The biological oxidation of ammonium under oxic conditions is already cumbersome; in addition, the harsh conditions required for the chemical oxidation of ammonium,

N. Perez (✉) · O. Martin · R. Cárdenas
Central University Marta Abreu from Las Villas, Santa Clara, VC, Cuba
e-mail: noelpd@uclv.edu.cu

O. Martin
e-mail: omgonzalez@uclv.cu

R. Cárdenas
e-mail: rcardenas@uclv.eedu.cu

R. Cárdenas et al. (eds.), *Proceedings of the 2nd International Conference on BioGeoSciences*, https://doi.org/10.1007/978-3-030-04233-2_7

very well known in industrial processes, seemed to be good reasons for the anammox process not to take place.

In 1977, Engelbert Broda predicted that the oxidation of ammonium under anoxic conditions with nitrite or nitrate as an electron acceptor "should exist or exist." Oxidation of ammonium with an electron acceptor other than oxygen was also previously predicted by marine environment researchers, based first on mass balance studies [1], later in combination with a thermodynamic justification [2]. These predictions were presented at a time when it was generally believed that the biological oxidation of ammonium without the presence of oxygen was simply impossible [3].

In 1985, the elimination of ammonium under anoxic conditions was first observed in a pilot-scale denitrification reactor at the yeast factory of the Gist-Brocades bakery in Delf, the Netherlands [4–7]. To achieve the oxidation of sulfur coupled to the reduction of nitrate, nitrate was added and, surprisingly, the elimination of ammonium in this reactor was also produced. This new biological process was called the anammox process. After the initial observations in Delf, in Germany [8, 9] and Switzerland [10], the production of dinitrogen gas instead of nitrate (nitrogen losses) in treated wastewater (rich in ammonia) from leachate was reported of landfills.

Since the second half of the nineteenth century, the biological conversion of nitrogen has had a marked scientific interest. The combination of the reactions of nitrogen fixation (conversion of dinitrogen to ammonium), nitrification (oxidation of ammonium to nitrate), and denitrification (reduction of nitrate to gaseous dinitrogen) allowed to interpret the conversions of nitrogen in nature as a biological cycle (Fig. 1). Based on the above, the cycle was formed that expressed the general view on nitrogen conversions during most of the twentieth century [3].

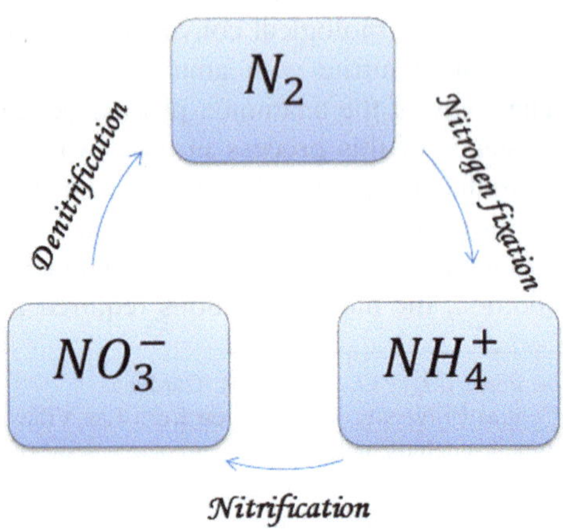

Fig. 1 Classical nitrogen cycle during twentieth century

2 Conventional Nitrification–Denitrification Processes

2.1 Nitrification

Nitrification is aerobic chemoautotrophic oxidation of inorganic nitrogenous compounds, such as (NH_4^+), hydroxylamine (NH_2OH), and nitrite (NO_2^-) to nitrate (NO_3^-) [11]. Complete nitrification takes place in two stages: the first being the oxidation of NH_4^+ to NO_2^-, and the second the nitrite becomes NO_3^-. Each conversion is carried out by different genera of bacteria. Oxidation of ammonia is carried out by Nitrosomonas, Nitrosococcus, Nitrosospira, Nitrosovibrio, and Nitrosobolus [11]. Such bacteria are also called ammonia-oxidizing bacteria, with hydroxylamine as an intermediate reaction [12]. Considering a cellular yield of 0.15 g cells/g NH_4^+–N, the equation for the oxidation of ammonia by Nitrosomonas [13] is as follows:

$$\begin{aligned} &NH_4^+ + 1.3818O_2 + 0.0909HCO_3^- \\ &\quad \rightarrow 0.0182C_5H_7NO_2 + 0.9818NO_2^- + 1.0364H_2O + 1.89H_2CO_3 \end{aligned} \tag{1}$$

The oxidation of nitrite to nitrate is carried out by a diverse group of proteobacteria called nitrite-oxidizing bacteria, among which are Nitrospira, Nitrospina, Nitrosococcus, and Nitrocystis, although Nitrobacter is the most recognized genus [11]. For a cellular yield of 0.02 g cells/g NO_2^--N, the metabolism of nitrite in Nitrobacter follows the reaction:

$$\begin{aligned} &NO_2^- + 0.0025NH_4^+ + 0.01HCO_3^- + 0.01H^+ + 0.048750O_2 \\ &\quad \rightarrow 0.0025C_5H_7NO_2 + 0.0075H_2O + NO_3^- \end{aligned} \tag{2}$$

By addition of the two processes including cell synthesis, the overall equation is represented as follows:

$$\begin{aligned} &NH_4^+ + 1.83O_2 + 1.98HCO_3^- \\ &\quad \rightarrow 0.021C_5H_7NO_3 + 0.98NO_3^- + 1.041H_2O + 1.88H_2CO_3 \end{aligned} \tag{3}$$

Nitrification is pH dependent, with the activity decreasing below pH 7.0, and it also consumes alkalinity, which is why it is commonly supplemented with $CaCO_3$ to avoid inhibition by pH [14]. The typical time of retention of sludge in activated sludge systems is between 10 and 20 days [14], due to the low rate of growth of microorganisms.

2.2 Denitrification

Denitrification is the biological reduction of NO_3^- and NO_2^- to N_2 gas. This is usually carried out along a heterotrophic process under anoxic–anaerobic conditions [11]. The complete reduction implies successive reductions with NO_2^-, nitric oxide (NO), and nitrous oxide (N_2O) as intermediates and can be performed by a single type of bacteria. To obtain energy, the different Gram-negative proteobacteria can use a wide range of organic compounds as electron donors and carbon source, as well as nitrates and nitrites as electron acceptors, producing nitrogen gas as the main product. Denitrifying bacteria occupy very diverse niches due to the great metabolic diversity existing between them. Assuming a cellular yield of 0.45 g cells/g NO_3^--N, with methanol as an electron donor [13], the alkalinity is generated in the process as shown in the following equation:

$$\begin{aligned} &NO_3^- + 1.08CH_3OH + 0.24H_2CO_3 \\ &\quad \rightarrow 0.056C_5H_7NO_2 + 0.47N_2 + 1.68H_2O + HCO_3^- \end{aligned} \tag{4}$$

Some examples of heterotrophic denitrifiers are: *Pseudomonas, Alcaligenes, Paracoccus, and Thiobacilus.*

2.3 Anaerobic Ammonium Oxidation

The anaerobic oxidation of NH_4^+ (anammox) with NO_2^- as terminal electron acceptor (catabolic reaction), yielding N_2 and NO_3^- as main products, is the recently discovered missing link in the N-cycle [15]:

$$NH_4^+ + NO_3^- \rightarrow N_2 + 2H_2O \quad \Delta G = -359\,kJ\Delta(mol\ NH_4^+)^{-1} \tag{5}$$

Several chemoautotrophic bacteria related to the genus planctomycetes carry out this process. Fundamentally, five genera "Candidatus," Brocadia [16], Kuenenia [17], Scalindua [18], Anammoxoglobus [19], and Jettenia [20] have been studied. Anammox bacteria are found in diverse habitats among them may be mentioned: marine sediments [21, 22], freshwater ecosystems [23, 24], and wastewater treatment plants [7]. These bacteria are considered responsible for up to 50% of oceanic *N* losses [25].

The growth rate of anammox bacteria is low oscillating the doubling time between 11 and 30 days [26, 27]. However, some specific studies have reported duplication times as low as 1.8 days [28] and 3 days [29]. The existence of low growth rates leads to low cellular performance as evidenced by fallow stoichiometry [27]:

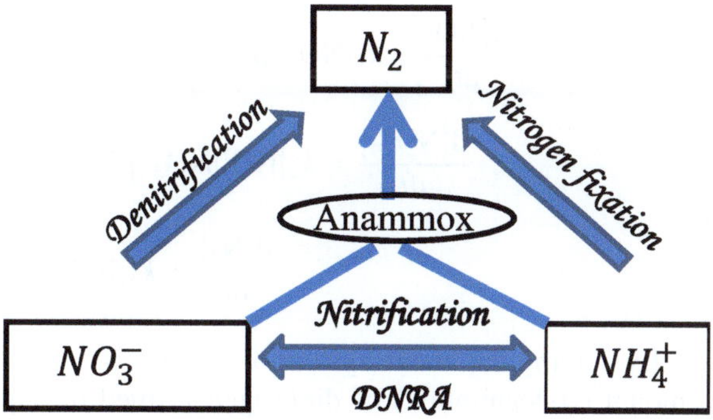

Fig. 2 Current microbial nitrogen cycle (DNRA: Dissimilatory nitrate reduction to ammonium)

$$\begin{aligned} & NH_4^+ + 1.32NO_2^- + 0.066HCO^- + 0.13H^+ \\ & \quad \rightarrow 1.02N_2 + 0.26NO_3^- + 0.066CH_2O_{0.5}N_{0.15} + 2.03N_2O \end{aligned} \tag{6}$$

The anammox process is used in an emergent way in the treatment of *N*-rich wastewater, especially in those where there is a low *C*/*N* ratio [30], mainly due to the high specific activity of around 0.8 kg N kg of dry weight $^{-1}$ day^{-1}.

Figure 2 shows a more complete picture of the nitrogen cycle in nature after the discovery of the anammox bacteria [30].

3 Kinetic of Anammox Catabolism

In the absence of molecular oxygen, it is very difficult to activate ammonium. For a long time; since the discovery of anammox bacteria, researchers have asked how is it possible that these microorganisms can oxidize ammonium along with the reduction of nitrite to form an N–N bond and produce nitrogen gas. Based on the in silico analysis of the genome assembly of the anammox bacterium Kuenenia stuttgartiensis, a set of three redox reactions (Eqs. 7–9) involving hydrazine (N_2H_4) and nitric oxide (NO) were proposed as intermediates to explain the stoichiometry of global anammox (Eq. 5) [15, 31]:

$$NO_2^- + 2H^+ + e^- \rightarrow NO + H_2O \quad k_1 \tag{7}$$

$$NO + NH_4^+ + 2H^+ + 3e^- \rightarrow N_2H_4 + H_2O \quad k_2 \tag{8}$$

$$N_2H_4 \rightarrow N_2 + 4H^+ + 4e^- \quad k_3 \tag{9}$$

In the above equations, k_1, k_2 and k_3 are the velocities' constants. Using formal kinetics, it is possible to obtain a system of three equations for the reaction rates of each stage that integrate the process described by Eq. (5), as shown follows:

$$\frac{d[NO]}{dt} = k_1[NO_2^-][H^+]^2[e] - k_2[NO][NH_4^+][H^+]^2[e]^3 \tag{10}$$

$$\frac{d[N_2H_4]}{dt} = k_2[NO][NH_4^+][H^+]^2[e]^3 - k_3[N_2H_4] \tag{11}$$

$$\frac{d[N_2]}{dt} = k_3[N_2H_4] \tag{12}$$

By solving the above system of equations, it is possible to roughly estimate the amount (concentration) of dinitrogen formed during the metabolism of anammox bacteria in any time. Of course, for a more realistic solution, it necessary to consider the inhibitory effect of some subtracts (nitrite and ammonium) or other species presents (hydroxylamine, hydrogen sulfur, dioxygen) and pH of medium, and introducing a term that represents the efficiency of the species. Considering the hypothesis of the steady state could simplify the problem but perhaps run the risk of being a bit unreal. Either way, these are lines of work for the future, which could be extended to other chemosynthetic autotrophs, as, for example, those that live in hydrothermal vents at great depths in the ocean that take advantage of hydrogen sulfide as a substrate.

4 Anammox and Chicxulub Impact

In the Black Sea and in general in all modern anoxic basins, ammonium is the only form of *N*-nutrients in the deep water below the chemocline, which diffuses upwards from the anoxic zone and is effectively consumed by the anammox bacteria before reaching the oxic zone [18]. The anammox bacteria are able to produce a loss of around 40% of the fixed nitrogen that sinks in the anoxic water of this sea. A significant increase in the areal extent of true anoxic conditions in the ocean, most likely could result in further extensive loss of N-nutrients through the anaerobic oxidation of ammonium during the Cretaceous oceanic anoxic event [32].

Most of the paleo-episodes of ocean acidification were too slow or too small as to be instructive in predicting impacts in the near future. The end-Cretaceous event (66 Mya) is intriguing in this regard, both because of its rapid onset and also because many pelagic calcifying species (including 100% of ammonites and more than 90% of calcareous nannoplankton and foraminifera) went extinct at this time. Sulfuric acid addition could have made the surface ocean extremely undersaturated for calcite, but only if they reached the ocean very rapidly (over few days) and if the quantity added was at the top end of the literature estimates [33, 34].

The impact of the Chicxulub asteroid produced at the end of the Cretaceous induced very sudden changes in atmospheric composition, climatic and planetary biodiversity. There was an abrupt and sudden acidification event fundamentally concentrated in the surface waters since the deep waters undergo a delayed and less severe acidification in response to an atmospheric source of acidity. Paleolithic records with which to restrict the changes produced in water chemistry of sea during those few post impact critical years after impact do not exist, and the low rate of accumulation of ocean sediments limit the resolution of sediments records to thousands of years. Tyrrell et al. in their work "Severity of ocean acidification following the end-Cretaceous asteroid impact" used models to calculate how dramatic the acidification of the surface ocean could have been at the end of the Cretaceous [33, 34].

Life on Earth and most likely the biospheres in exoplanets are seriously threatened by the impacts of asteroids and comets. Mass extinction in the Cretaceous–Paleogene (K/Pg) boundary where dinosaurs practically disappeared and approximately 50% of the living genera, it widely accepts as the main contributor the impact of the Chicxulub asteroid. Because of this natural catastrophe, there were several environmental stresses. The most accepted scenario immediately after the impact is "cold and dark." The aerosols (mainly sulfate aerosols), soot, and dust in the atmosphere completely covered the sunlight at least during half a year (some researchers estimate approximately two and a half years), with the consequent collapse of photosynthesis and global deforestation. The ozone layer was completely destroyed, due to the release of large amounts of chlorine and bromine that form the evaporation of both the asteroid and the target rocks [35]. The target rocks formed mainly by sediments rich in carbonates and gypsum/anhydrite with a granite crust were partially ejected and volatilized as a result of the strong impact. In addition to the sulfur generated by target rock, about $1–5 \times 10^{15}$ mol came from the asteroid itself. The thermal decomposition of gypsum or anhydrite, on the basis of the experiments of volatilization carried out in laboratories, could lead to the almost instantaneous release of sulfur trioxide (SO_3) into the atmosphere according to the reaction:

$$CaSO_4 \rightarrow CaO + SO_3 \quad (13)$$

Once the sulfur trioxide was injected into the atmosphere, it could have been transformed into sulfuric acid by reacting with the water, which would later fall on the ocean in the form of acid rain. Several atmospheric modeling of the K/Pg limit [36, 37] used atmospheric residence times of sulfur from several months to a few years. However, recent studies have proposed an alternative scenario and very different from the one generally accepted for the K/Pg event in which it is suggested [38] that, immediately after the impact, most of the sulfate aerosols (sulfuric acid aerosols) could have been eliminated by the large particles of silicates that fell rapidly back to Earth, delivering the load of H_2SO_4 to the ocean in only one or a few days. Ocean acidification may have affected the metabolism of anammox bacteria, probably leading to inhibition, which should be reflected in their activity. Tyrrell et al. [33] considered additions of sulfuric acid of 15, 30, and 60×10^{15} mol, corresponding to 480, 960, and 1920 Pg, with e-folding timescale from 10 h to 5 years to have a range of possibilities more large. The addition of sulfuric acid reduces the total alkalinity (TA) of the ocean surface water in a molar ratio H_2SO_4 : TA = 1 : 2 [39].

Similar to the production of sulfur compounds due to the vaporization of gypsum rocks, the vaporization of carbonate rocks produces CO_2. The Chicxulub asteroid with approximately a diameter of 10 km hits a layer of sedimentary carbonates of about 3–4 km thick in the Yucatan peninsula [40] releasing between 5000 and 9000 Pg of CO_2 [41] (equivalent to 1300 and 2500 Pg of carbon, respectively). This calculation can be overestimated because possibly the largest amount of carbon dioxide released was greatly reduced due to the rapid recombination [42] of volatilized CaO (around 40–80%) and CO_2 within the impact plume to form calcium carbonate again [43]. After the impact, all the woody biomass could have ignited due to a global thermal shock, which caused forest fires [44, 45] and therefore release large amounts of CO_2 into the atmosphere. It is unlikely that the combustion of carbon from terrestrial vegetation could have contributed much more than 1500 Pg of carbon, both by the aridity of the mid-continent in warmer climates and by the finite habitat space [33]. On the other hand, another potential source of carbon comes from soils. Because of the impact, large amounts of soot, dust, and aerosols were released into the atmosphere, bringing a long period of darkness on the surface of the Earth, during which the process of photosynthesis was strongly inhibited due to low levels of light available [46]. The decomposition of organic carbon (with a rotation time currently of approximately 50 years) [47] could not be balanced with the replacement of the production of leaf litter and other carbon-rich material by living plants. Tyrrell et al. assumed a maximum total for the Upper Cretaceous of 2500 Pg C (1600 Pg at present) considering that the carbon reserves of the soil in the Earth at present are much higher toward the poles, particularly in the permafrost regions. Although the Earth was warmer in the late Cretaceous, the lack of ice in Antarctica could have allowed the accumulation of large carbon reserves in the soil [48]. Tyrrell et al. modeled the effect on the oceanic carbonate chemistry of all these combined sources considering carbon additions of 2000, 4000, and 6500 Pg of C. The volatilization of the carbonate rock in the impact region plus the forest fires and the decay of the soil carbon were fast and slow sources of carbon release, respectively. According to the study by Tyrrell et al. [33], large impacts are produced but not as severe as from large sulfuric additions. In contrast to H_2SO_4, both slower and faster additions of CO_2 to the atmosphere cause similar responses in the ocean, because the slow air–sea exchanges of CO_2 delay the onset of ocean acidity.

The asteroid (and the subsequent ejection) when crossing the atmosphere did so at high speed and the intense wave pressure associated with it should have favored the formation of NO_x from the atmospheric N_2 and O_2. When NO_x reacts with water present in the atmosphere, it forms nitric acid (HNO_3) which, when incorporated into the rain, can induce ocean acidification (similar to what occurs with SO_3) in the following months or years (or possibly days if it is scavenged by the large silicate particles). The total amount of HNO_3 is generated mainly by three causes: direct production of the initial pressure wave (1×10^{15} mol), pressure ejection (double the amount produced by the previous route probably), and forest fires

(3×10^{15} mol). The three previous routes result in a maximum ascent of 5 × 1015 mol of HNO_3. Tyrrell et al. [33] considered the additions of HNO_3 of 1, 3, and 5×10^{15} mol in their study. The impact is similar to that of H_2SO_4, but considerably smaller.

According to the study by Tyrrell et al. [33], changes in ocean pH are mainly due to the injection of large amounts of H_2SO_4 more than those of HNO_3 and CO_2. The more quickly H_2SO_4 is added to the ocean, the more intense the changes in pH are. Therefore, it is natural that ocean acidification could affect the activity of bacteria, and in particular anammox.

In several studies on anammox, the so-called anammox activity (AA) is used, which is measured according to the production rate of N_2 and is expressed as mmol $N_2\,L_{\text{liquid}}^{-1}\,h^{-1}$, as indicated the equation that follows:

$$AA = \frac{\Delta N_2}{\Delta t} \tag{14}$$

The inhibition was expressed as relative activity of anammox in percent (RAA) by:

$$RAA = \frac{AA_{\text{inhibitor}}}{AA_{\text{reference}}} \times 100 \tag{15}$$

In the above expression, $AA_{\text{reference}}$ is the value of AA to optimal pH and $AA_{\text{inhibitor}}$ is the value of AA at different pH [13]. The effect on RAA associated with the changes of the pH of the ocean is shown in Fig. 3. Inhibition is larger the higher the change in the pH of the ocean and lasts longer in time as the addition of H_2SO_4 is greater.

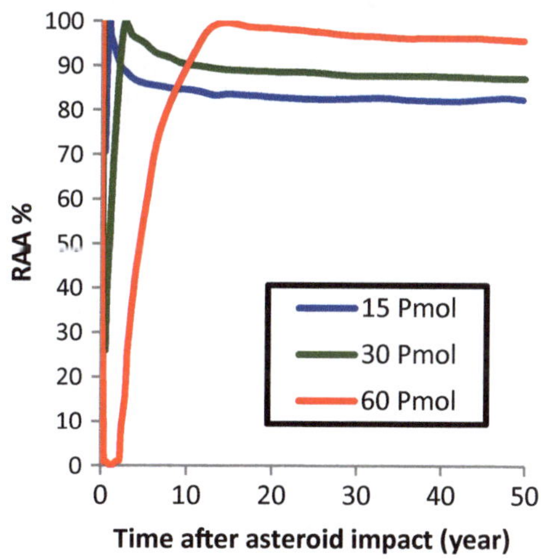

Fig. 3 Effect of pH on relative activity of anammox (RAA) on suspended culture of anammox. Time e-folding of 6 months

It is to be expected that with rapid additions (smaller timescale), the inhibitory effects on anammox are more abrupt, probably for all H_2SO_4 inputs, but this would be the subject of future work.

The impact of a comet and intense magmatic events/volcanism can generate similar effects on anammox bacteria (and in general, on other chemosynthetic species). These catastrophic events not only affect photosynthetic life as seen in other works, but also the chemosynthetic biota and therefore the habitability of ecosystems and the planet in general.

References

1. Richards F (1965) Anoxic basins and fjords. In: Riley JP, Skirrow G (eds) Chemical oceanography, vol 1. Academic Press, New York, pp 611–641
2. Cline, J.D and Richards, F.A.: Oxygen deficient conditions and nitrate reduction in the eastern tropical North Pacific Ocean. Limnol Oceanogr **17**(6) 885–900 (1972)
3. Van der Star W (2008) Growth and metabolism of anammox bacteria. biofilms.bt.tudelft.nl/pdf/EBTprVdStar.pdf
4. Heijnen JJ (1988) Biologische anaëroob-aërobe afvalwaterzuivering bij Gist-Brocades: eindrapport 1977–1986. 's-Gravenhage, NL, Staatsuitgeverij/DOP. ISBN 978 90 346 1686 X
5. Mulder A (1989) Anoxic ammonia oxidation of wastewater. European Patent Ep327184. Assignee: Gist-Brocades NV, NL
6. Van de Graaf AA, Mulder A, Slijkhuis H, Robertson LA, Kuenen JG (1990) Anoxic ammonium oxidation. Proc Eur 5th Congr Biotechnol I:388–391
7. Mulder A, Van de Graaf AA, Robertson LA, Kuenen JG (1995) Anaerobic ammonium oxidation discovered in a denitrifying fluidized bed reactor. FEMS Microbiol Ecol 16(3): 177–184
8. Hippen A, Rosenwinkel KH, Baumgarten G, Seyfried CF (1996) Aerobic deammonification: a new experience in the treatment of wastewaters. Mededelingen - Faculteit Landbouwkundige en Toegepaste Biologische Wetenschappen (Universiteit Gent) **61**(4b) 1967–1974
9. Hippen A, Rosenwinkel KH, Baumgarten G, Seyfried CF (1997) Aerobic de-ammonification: anew experience in the treatment of wastewaters. Water Sci Technol 35(10):111–120
10. Binswanger S, Siegrist H, Lais P (1997) Simultane Nitrifikation/Denitrifikation von stark ammoniumbelasteten Abwassern ohne organische Kohlenstoffquellen [Simultaneous nitrification/denitrification of wastewaters polluted with high levels of ammonium in the absence of organic carbon sources]. Korresp Abwasser 44(9):1573–1580
11. Wong CH, Barton GW, Barford JP (2003) The nitrogen cycle and its application in wastewater treatment. Handbook of water and wastewater microbiology. Academic Press, London
12. Güven D, Schmidt I (2009) Specific activity and viability of *Nitrosomonas europaea* during discontinuous and continuous fermentation. Process Biochem 44(5):516–520
13. Carvajal-Arroyo JM (2013) Inhibitory impact of nitrite on anaerobic ammonium oxidizing (anammox) bacteria: inhibition mechanisms and strategies to improve the reliability of the anammox process as a N-removal technology. University Libraries, The University of Arisona. http://hdl.handle.net/10150/311350
14. Metcalf E, Tchobanoglous G, Burton FL, Stensel HD (2003) Wastewater engineering: treatment and reuse. McGraw-Hill, Boston
15. Strous M, Pelletier E, Mangenot S, Rattei T, Lehner A, Taylor MW, Horn M, Daims H, Bartol-Mavel D, Wincker P, Barbe V, Fonknechten N, Vallenet D, Segurens B, Schenowitz-Truong C, Medigue C, Collingro A, Snel B, Dutilh BE, Op den Camp HJM,

van der Drift C, Cirpus I, van de Pas-Schoonen KT, Harhangi HR, van Niftrik L, Schmid M, Keltjens J, van de Vossenberg J, Kartal B, Meier H, Frishman D, Huynen MA, Mewes HW, Weissenbach J, Jetten MSM, Wagner M, Le Paslier D (2006) Deciphering the evolution and metabolism of an anammox bacterium from a community genome. Nature 440(7085): 790–794

16. Strous M, Fuerst JA, Kramer EHM, Logemann S, Muyzer G, van de Pas-Schoonen KT, Webb R, Kuenen JG, Jetten MSM (1999) Missing lithotroph identified as new planctomycete. Nature 400(6743):446–449
17. Schmid M, Schmitz-Esser S, Jetten M, Wagner M (2001) 16S-23S rDNA intergenic spacer and 23S rDNA of anaerobic ammonium-oxidizing bacteria: implications for phylogeny and in situ detection. Environ Microbiol 3(7):450–459
18. Kuypers MMM, Sliekers AO, Lavik G, Schmid M, Jorgensen BB, Kuenen JG, Damste JSS, Strous M, Jetten MSM (2003) Anaerobic ammonium oxidation by anammox bacteria in the Black Sea. Nature 422(6932):608–611
19. Kartal B, Rattray J, van Niftrik LA, van de Vossenberg J, Schmid MC, Webb RI, Schouten S, Fuerst JA, Damste JSS, Jetten MSM, Strous M (2007) Candidatus "*Anammoxoglobus propionicus*" a new propionate oxidizing species of anaerobic ammonium oxidizing bacteria. Syst Appl Microbiol 30(1):39–49
20. Quan ZX, Rhee SK, Zuo JE, Yang Y, Bae JW, Park JR, Lee ST, Park YH (2008) Diversity of ammonium-oxidizing bacteria in a granular sludge anaerobic ammonium-oxidizing (anammox) reactor. Environ Microbiol 10(11):3130–3139
21. Schmid MC, Risgaard-Petersen N, van de Vossenberg J, Kuypers MMM, Lavik G, Petersen J, Hulth S, Thamdrup B, Canfield D, Dalsgaard T, Rysgaard S, Sejr MK, Strous M, den Camp HJMO, Jetten MSM (2007) Anaerobic ammonium-oxidizing bacteria in marine environments: widespread occurrence but low diversity. Environ Microbiol 9(6):1476–1484
22. Rich J, Dale O, Song B, Ward B (2008) Anaerobic ammonium oxidation (Anammox) in Chesapeake Bay sediments. Microb Ecol 55(2):311–320
23. Schubert CJ, Durisch-Kaiser E, Wehrli B, Thamdrup B, Lam P, Kuypers MMM (2006) Anaerobic ammonium oxidation in a tropical freshwater system (Lake Tanganyika). Environ Microbiol 8(10):1857–1863
24. Zhang Y, Ruan X-H, den Camp HJMO, Smits TJM, Jetten MSM, Schmid MC (2007) Diversity and abundance of aerobic and anaerobic ammonium-oxidizing bacteria in freshwater sediments of the Xinyi River (China). Environ Microbiol 9(9):2375–2382
25. Kuypers MMM, Lavik G, Woebken D, Schmid M, Fuchs BM, Amann R, Jørgensen BB, Jetten MSM (2005) Massive nitrogen loss from the Benguela upwelling system through anaerobic ammonium oxidation. Proc Natl Acad Sci USA 102(18):6478–6483
26. Van de Graaf AA, de Bruijn P, Robertson LA, Jetten MSM, Kuenen JG (1996) Autotrophic growth of anaerobic ammonium-oxidizing micro-organisms in a fluidized bed reactor. Microbiology 142(8):2187–2196
27. Strous M, Heijnen JJ, Kuenen JG, Jetten MSM (1996) The sequencing batch reactor as a powerful tool for the study of slowly growing anaerobic ammonium-oxidizing microorganisms. Appl Microbiol Biotechnol 50(5):589–596
28. Isaka K, Date Y, Sumino T, Yoshie S, Tsuneda S (2006) Growth characteristic of anaerobic ammonium-oxidizing bacteria in an anaerobic biological filtrated reactor. Appl Microbiol Biotechnol 70(1):47–52
29. Van der Star WRL, Miclea AI, van Dongen U, Muyzer G, Picioreanu C, van Loosdrecht MCM (2008) The membrane bioreactor: A novel tool to grow anammox bacteria as free cells. Biotechnol Bioeng 101(2):286–294
30. Kartal B, van Niftrik L, Sliekers O, Schmid MC, Schmidt I, van de Pas-Schoonen K, Cirpus I, van der Star W, van Loosdrecht M, Abma W, Kuenen JG, Mulder J-W, Jetten MSM, den Camp HO, Strous M, van de Vossenberg J (2004) Application, eco-physiology and biodiversity of anaerobic ammonium-oxidizing bacteria. Rev Environ Sci Biotechnol 3 (3):255–264

31. Kartal B, Maalcke WJ, de Almeida NM, Cirpus I, Gloerich J, Geerts W, Op den Camp HJM, Harhangi HR, Janssen-Megens EM, Francoijs KJ, Stunnenberg HG, Keltjens JT, Jetten MS, Strous M (2011) Molecular mechanism of anaerobic ammonium oxidation. Nature 479: 127–130
32. Kuypers MMM, van Breugel Y, Schouten S, Erba E, Damsté JSS (2004) N_2-fixing cyanobacteria supplied nutrient N for Cretaceous oceanic anoxic events. Geology 32:853–856
33. Tyrrell T, Merico A, Armstrong M, Kay DI (2015) Severity of ocean acidification following the end-Cretaceous asteroid impact. PNAS 112:6556–6561
34. Tyrrell T, Merico A, Armstrong M, Kay DI (2015) Supporting information online at http://www.pnas.org/lookup/suppl/, https://doi.org/10.1073/pnas.1418604112/-/dcsupplemental
35. Perez N, Cardenas R, Martin O, Rojas R (2013) Modeling the onset of fotosynthesis after the Chicxulub asteroid impact. Astrophys Space Sci 343:7–10
36. Pierazzo E, Hahmann AN, Sloan LC (2003) Chicxulub and climate: Radiative perturbations of impact-produced S-bearing gases. Astrobiology 3(1):99–118
37. Pope KO, Baines KH, Ocampo AC, Ivanov BA (1994) Impact winter and the Cretaceous/Tertiary extinctions: results of a Chicxulub asteroid impact model. Earth Planet Sci Lett 128 (3):719–725
38. Ohno S, Kadono T, Kurosawa K, Hamura T, Sakaiya T, Shigemori K, Hironaka Y, Sano T, Watari T, Otani K, Matsui T, Sugita S (2014) Production of sulphate-rich vapour during the Chicxulub impact and implications for ocean acidification. Nat Geosci 7(4):279–282
39. Doney SC et al (2007) Impact of anthropogenic atmospheric nitrogen and sulfur deposition on ocean acidification and the inorganic carbon system. Proc Natl Acad Sci USA 104 (37):14580–14585
40. Schulte P et al (2010) The Chicxulub asteroid impact and mass extinction at the Cretaceous-Paleogene boundary. Science 327(5970):1214–1218
41. O'Keefe JD, Ahrens TJ (1989) Impact production of CO_2 by the Cretaceous Tertiary extinction bolide and the resultant heating of the earth. Nature 338(6212):247–249
42. Yancey TE, Guillemette RN (2008) Carbonate accretionary lapilli in distal deposits of the Chicxulub impact event. Geol Soc Am Bull 120(9–10):1105–1118
43. Agrinier P, Deutsch A, Schärer U, Martinez I (2001) Fast back-reactions of shock-released CO2 from carbonates: An experimental approach. Geochim Cosmochim Acta 65(15): 2615–2632
44. Goldin TJ, Melosh HJ (2009) Self-shielding of thermal radiation by Chicxulub impact ejecta: firestorm or fizzle? Geology 37(12):1135–1138
45. Morgan J, Artemieva N, Goldin T (2013) Revisiting wildfires at the K-Pg boundary. J Geophys Res-Biogeo 118(4):1508–1520
46. Vellekoop J et al (2014) Rapid short-term cooling following the Chicxulub impact at the Cretaceous-Paleogene boundary. Proc Natl Acad Sci USA 111(21):7537–7541
47. Giardina CP, Ryan MG (2000) Evidence that decomposition rates of organic carbon in mineral soil do not vary with temperature. Nature 404(6780):858–861
48. DeConto RM et al (2012) Past extreme warming events linked to massive carbon release from thawing permafrost. Nature 484(7392):87–91

Quantification of Phytoplankton Primary Habitability in the Gulf of Ana María, Cuba

Jessica Alvarez-Salgueiro, Dailé Avila-Alonso, Rolando Cárdenas, Roberto González-De Zayas and Osmel Martin

Abstract The Gulf of Ana Maria is located in the southeast Cuban platform and is one of the most important regions for fisheries in the country. Taking into account that marine primary production defines the areas and yields of fishery production, we develop a habitability index that quantifies the environmental suitability for phytoplankton assemblage development. The index considers light and limiting nutrient as the main environmental factors governing the phytoplankton primary production in the Gulf of Ana María. We considered a function of radiation described by a photosynthesis model and function of nutrient described by a eutrophication index. Quantitative estimates of habitability were made in nine stations at 1:00 PM and 4:00 PM for a typical day of October, in order to match with the available field data. Solar irradiation at sea surface was calculated by using the tropospheric ultraviolet and visible computational code TUV4.6, and we derived the data of attenuation coefficient of light from Secchi depth data. Besides, we used nutrient concentration of soluble reactive phosphorus (limiting nutrient). We obtained the first optical classification of the waters of the Gulf of Ana María, with a dominance of coastal water C5. It was found an intermediate potential ($\sim$50%) for phytoplankton development in the Gulf of Ana María, which is mainly limited by nutrients rather than by light. Further refinements in both the index and data collection are expected to reach more accurate conclusions.

Keywords Habitability index · Nutrients · Light

J. Alvarez-Salgueiro
Universidad de Camagüey Ignacio Agramonte y Loynaz, Bypass Road, km 5.5, Camagüey, Cuba

D. Avila-Alonso · R. Cárdenas (✉) · O. Martin
Universidad Central "Marta Abreu" de Las Villas, Camajuaní Road, km 5.5, Santa Clara, Villa Clara, Cuba
e-mail: rcardenas@uclv.edu.cu

R. González-De Zayas
Coastal Ecosystem Research Center (CIEC), Los Almácigos Avenue, Cayo Coco, Morón, Ciego de Ávila, Cuba

R. Cárdenas et al. (eds.), *Proceedings of the 2nd International Conference on BioGeoSciences*, https://doi.org/10.1007/978-3-030-04233-2_8

1 Introduction

Coastal environments are one of the most ecologically and socioeconomically important on the planet. Despite their relatively small coverage (occupying only ~8% of the ocean's surface area), they contribute to 15–30% of global primary production, >40% of seabed carbon sequestration, and 80% of world fishing capture [1–3]. Coastal ecosystems have an important role in the global biogeochemical cycles of carbon and macronutrients [4], and they are considered buffering systems between the continents and the open ocean. The coastal ocean provides nutrients to the open ocean, and much of these nutrients come from rivers runoff [5]. Some quantitative studies have shown that when the coastal nutrient supply is enhanced, both the coastal ocean and the open ocean primary production and chlorophyll-a (chl-a) concentration increase in comparable amounts [6]. However, in general, coastal phytoplankton assemblages are more sensitive to environmental changes than the oceanic ones [1, 2].

The coastal waters of Cuba are very important for the biogeographic processes affecting the biodiversity of the so-called Great Caribbean Region and for conservation and sustainable use of natural resources in this region [7]. Moreover, it has been reported that several regions of the Cuban coasts fertilized the oligotrophic waters of the Caribbean Sea [8]. However, coastal waters of Cuba are threatened by fishing, tourism, as well as other economic activities associated with the marine platform [7]. Especially, phytoplankton is very vulnerable in some regions of the country due to both natural variability and anthropogenic activity [9].

The main flux of nutrients into the Cuban coastal waters comes from hydrographic basins; thus, damming of rivers has recently caused a reduction of nutrient loading to coastal waters [10, 11]. This situation has affected the Gulf of Ana María (located in southeastern Cuban shelf), whose hydrochemical conditions are characteristics of a coastal zone with excellent interchange of nutrients with the adjacent Caribbean Sea [12]. On the other hand, this region is one of the main fishing areas of the country, so that the productivity of various fisheries resources has declined, such as the shrimp species *Farfantepeneaus notialis* [11, 13–15]. A reduced flow of freshwater into the coastal waters causes a large variability of nutrients concentrations in the area which can affect the phytoplankton photosynthesis rate and the distribution and abundance of organisms of higher levels of the trophic webs. In this way, the potential for life in the ecosystem (i.e., its habitability) could change.

The Quantitative Habitability Theory (QHT) has been developed as a standard framework to quantify changes in the living conditions of a natural environment and to explain the distribution, abundance, and productivity of life [16, 17]. For instance, the Standard Primary Habitability (SPH) is a climatological measure of habitability of a region for most primary producers. It is a normalized scale, where values close to one represent the best environmental conditions for the metabolic activity of most organisms, while values close to zero are unfavorable [16].

The abundance and distribution of phytoplankton can be determined by several variables such as nutrients, salinity, weather conditions (light, temperature, rainfall, etc.), and the shape of aquatic system (depth, volume, surface area, etc.) [18]. However, [19] considered that solar radiation is a factor of first-order driving photosynthetic activity in marine ecosystems, while nutrient concentration is a second-order factor. Given the estuarine conditions of the Gulf of Ana María, it could be though that salinity has an important role in influencing primary productivity of phytoplankton assemblages. However, spatiotemporal variability of salinity in this area has been reported to be limited. For instance, it has been demonstrated that salinity was rather homogenous in 38 stations of the Ana María Gulf, during the raining season (October of 2011) with values in the range 36.0–37.3 ups [12]. Besides, in [20], it is reported similar values of salinity (i.e., 36.46–38.31 ups) along the coastal line of the Gulf of Ana María during the transition between dry and raining seasons and during rainy months.

Taking into account the situation previously described, this work aims to determine the habitability potential of phytoplankton in the Gulf of Ana María, through the construction of an Aquatic Primary Habitability Index (APHI), considering light and nutrients as possible environmental regulatory factors of photosynthetic activity.

2 Materials and Methods

2.1 Study Area

The Gulf of Ana María is a big macro-lagoon located in the southern-central shelf of Cuba. It is limited to the north by the provinces of Sancti Spiritus, Ciego de Avila, and Camagüey and to the south by "Jardínes de la Reina" archipelago, which regulates the water flows with the adjacent Caribbean Sea. The Gulf of Ana María has a surface area of 9398 km^2 and an average (maximum) depth of 15 m (30 m) [21]. One of the most relevant characteristics of this geographic basin is determined by rivers, which periodically (although unevenly) contribute to the increase of dissolved nutrients and suspended particles concentration in its waters. The rivers that flow into the Gulf of Ana María are Zaza, Agabama, southern Jatibonico, and San Pedro [12].

2.2 Aquatic Primary Habitability Index

One of the main postulates of QHT is related to the formulation of habitability indexes (HI) as the product of simple functions (*f*) of a set of environmental variables ($\{x_j\}$) (Eq. 1) that limits the primary productivity of life at any environment [16]:

$$\mathrm{HI} = \prod_{i=1}^{n} f_i(\{x_j\}). \tag{1}$$

Thus, we propose an APHI of the form:

$$\mathrm{APHI} = f(R)f(N) \tag{2}$$

where $f(R)$ is a function of solar radiation and $f(N)$ is a function of nutrients.

Radiation function *f*(*R*) The photosynthetic potential of phytoplankton in nine stations in the Gulf of Ana María (Fig. 1) was quantified, and the sampling activities were developed in October of 2011. The stations were selected taking into account that the depth of the station was higher than the Secchi depth. The spectral irradiances (W m^{-2} nm^{-1}) of the photosynthetically active radiation (PAR, 400–700 nm) and ultraviolet (UV, 280–399 nm) at the sea surface were determined using the tropospheric ultraviolet and visible computational code TUV 4.6 [22] at 1:00 PM and 4:00 PM and considering solar zenith angles of incidence of 35° and 76°, respectively. These conditions match the sampling campaign.

The spectral irradiances just below the surface ($z = 0^-$), denoted by $E(\lambda, 0^-)$, were obtained from those just above the surface ($z = 0^+$) through the following equation:

$$E(\lambda, 0^-) = [1 - R]E(\lambda, 0^+), \tag{3}$$

where R is the reflection coefficient calculated with Fresnel formulae [23] applied to the interface air–water. The spectral irradiances $E(\lambda, z)$ at depth z were determined using Lambert–Beer's law of optics as:

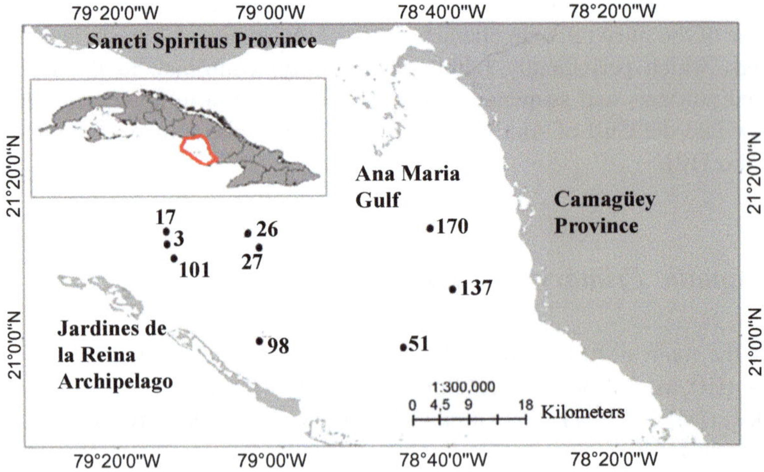

Fig. 1 Study area and sampling stations

$$E(\lambda, z) = E(\lambda, 0^-)e^{-K(\lambda)z}, \tag{4}$$

where $K(\lambda)$ is the light extinction coefficient or so-called attenuation coefficient for the wavelength λ. This parameter is a measure of the vertical light attenuation through the water column [18]. Given to the scarce published studies for the Gulf of Ana María ([12] and references therein), there is not an optical characterization of its marine waters, and consequently, there are not data of attenuation coefficients. In order to determine the optical type of water at each station (and therefore the set of values of $K(\lambda)$, the following empirical relation [24] was used:

$$K = \frac{k}{z_S}, \tag{5}$$

where K is the average attenuation coefficient of water for visible light (PAR), k is a dimensionless constant, which is considered here equal to 2 [24], and z_s is the Secchi depth. z_s together with nutrient concentration were obtained from a field campaign conducted during October 2011 [12]. Once the average attenuation coefficients were obtained through Eq. (5), they were compared with the coefficients of Jerlov [25] (Table 1). The optical water type assigned to each station was the one closer to Jerlov's value.

Beside the Secchi depth, another way to evaluate the underwater light penetration is to determine the euphotic depth (z_{EU}), which is relatively easy and accurate to estimate [26] by determining the depth at which 1% of the PAR entering the water remains. Making the assumption that K of PAR is approximately constant with depth [23], the euphotic depth is given by the following equation, which was obtained from the Lambert–Beer equation.

$$z_{EU} = \frac{4.6}{K} \tag{6}$$

Equations (7) and (8) were used to determine the irradiances of the photosynthetically usable radiation (PUR) and UV, respectively. The asterisks indicate that

Table 1 Average attenuation coefficients and optical classification of Jerlov for marine waters [25]

Waveband	Coastal waters					Oceanic waters		
	C1	C3	C5	C7	C9	I	II	III
UV	1.35	1.84	2.65	3.18	4.12	0.09	0.30	0.50
PAR	0.29	0.38	0.51	0.71	1.04	0.15	0.19	0.25
UV + PAR	0.59	0.79	1.12	1.41	1.92	0.14	0.21	0.31

both are photometric functions (i.e., they contain biological information through the inclusion of a biological action spectrum):

$$E^*_{\mathrm{PUR}}(z) = \sum_{\lambda=400\mathrm{nm}}^{700\mathrm{nm}} A(\lambda)E(\lambda, z)\Delta\lambda \tag{7}$$

$$E^*_{\mathrm{UV}}(z) = \sum_{\lambda=280\mathrm{nm}}^{399\mathrm{nm}} \varepsilon(\lambda)E(\lambda, z)\Delta\lambda. \tag{8}$$

In the first one, $A(\lambda)$ is the absorptance, the fraction of light absorbed by the cell [23] and $\Delta\lambda = 1$ nm. Considering that in general there are very few studies of phytoplankton assemblages in the Gulf of Ana María, we used the absorption spectrum of the chl-a. This one constitutes most of the chlorophyll present in primary producers [23]. Measurements of this photopigment historically have provided an estimate of algal biomass and its spatial and temporal variability [27]. Chl-a has two main absorption bands in the red and in the blue regions of the spectrum (Fig. 2).

To evaluate the inhibitory effects of UV on photosynthesis, we used the same biological weighting function ($\varepsilon(\lambda)$) used in [29] which is representative of phytoplankton species living in tropical and subtropical latitudes. This function takes into account the photosynthetic apparatus and DNA damages, since both contribute to the inhibition of photosynthesis.

To determine photosynthesis rates ($P(z)$) at depth z normalized with respect to the saturation rate P_S, the photosynthesis E model of [30] was used:

$$\frac{P(z)}{P_S} = \frac{1 - \exp[-E_{\mathrm{PAR}}(z)/E_S]}{1 + E^*_{\mathrm{UV}}(z)}. \tag{10}$$

The parameter E_S accounts for the efficiency with which a species uses PUR; the smaller its value, the greater its efficiency. Taking into account the limited number of studies on phytoplankton in the study area, we assumed, as our first approach, the values $E_S = 2$ W m^{-2} and $E_S = 100$ W m^{-2}, which represent a very wide range of this parameter in nature [31, 32].

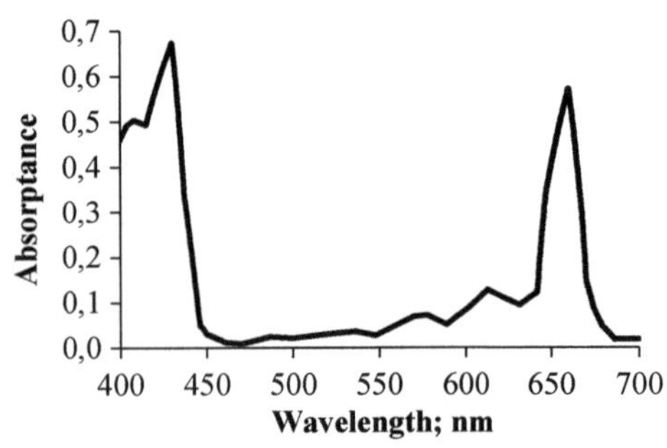

Fig. 2 Absorptance of the chl-a molecule [28]

The exact value of *f*(*R*) was determined through Eq. (11), which represents the total photosynthetic potential $(P/P\text{s})_\text{T}$, but normalized to the station depth $z_{\max}$. $(P/P\text{s})_\text{T}$ was calculated by splitting the water column into *N* layers of thickness $\Delta z = 1$ m:

$$f(R) = \left(\frac{P}{P_\text{S}}\right)_\text{T} = \frac{1}{z_{\max}} \sum_{i=1}^{N} \left(\frac{P}{P_\text{S}}\right)_i \Delta z, \tag{11}$$

where $(P/P\text{s})_i$ is the photosynthetic rate inside the *i*th layer of the water column.

Nutrient function *f*(*N*) As the function of (limiting) nutrient, we used the Eutrophication index proposed by [33]:

$$\text{EI} = \frac{C}{C - \log x_i} + \log A, \tag{12}$$

where EI is the eutrophication by the limiting nutrient in each sampling station for the sampling campaign, *A* is the number of sampling stations, *C* is the logarithm of the total concentration of limiting nutrient per campaign, and x_i is the total nutrient concentration in the sampled station. The grading scale is as follows: EI < 3 indicates *oligotrophic* conditions, ≤ 3 EI $\leq$ 5 *mesotrophic,* and EI > 5 *eutrophic*.

Taking into account the findings of [12], we considered the soluble reactive phosphorus (SRP) as the limiting nutrient in the waters of the Gulf of Ana María. The *APHI* was normalized with respect to the maximum value obtained at 1:00 PM and 4:00 PM, for each case. The spatial distribution of normalized *APHI* was interpolated by SURFER 9 package.

3 Results

3.1 Optical Water Classification

The values of the light attenuation coefficient for the waters of the Gulf of Ana María ranged from 0.31 to 0.73 m^{-1}, with an average value of 0.50 m^{-1}. The lowest (highest) attenuation coefficient value was found in the southwestern (southeastern) area of the gulf at station 3 (137) (Fig. 3). According to the modeling framework developed in this study, the marine waters of the Gulf of Ana María are categorized as coastal waters according to the marine water classification of Jerlov [25]. The clearest water type found was C1 (station 3), and the more turbid water type was C7 (station 137) (Fig. 3). There was not a representation of the darkest coastal waters (C9), while the most abundant optical water type was the intermediate turbidity (C5) (Fig. 3). Seawaters showed spatial variability in their optical properties. We identified three optical subregions in the Gulf of Ana María. The first one was located near the coastline, showing the darkest waters found in this study; the second subregion was located in the central gulf and is characterized by an intermediate turbidity, while the third subregion was at the western gulf and shows the clearest water types.

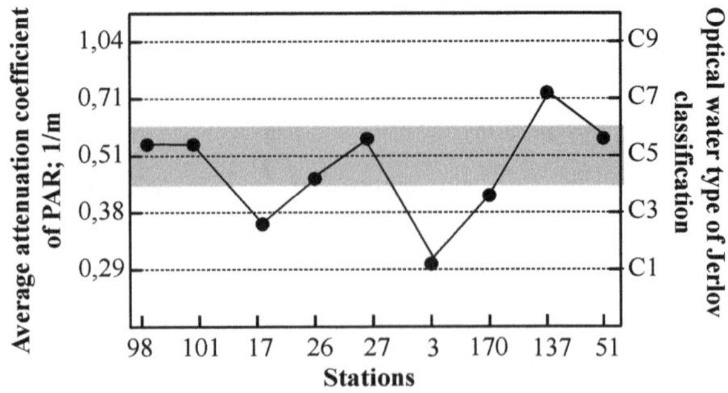

Fig. 3 Average attenuation coefficient of light and optical classification of the waters of the Gulf of Ana María. The gray band indicates stations with intermediate turbidity (C5 waters)

The average depth of the study area was 11 m, while the maximum and minimum depths were 21 m (station 101) and 5 m (station 137), respectively. The calculated extension of the euphotic layer, at most stations, was deeper than the station depth (Table 2), while the Secchi depths were relatively shallow (Table 2). For what concerns the trophic conditions, we found that all stations presented oligotrophic conditions, except the station 101 which showed eutrophic conditions (Table 2). However, we can observe that the values of the EI for SRP are relatively close to the lower limit of the mesotrophic classification according to [33] (see Sect. 2). This result could be related to the dominance of waters with intermediate turbidity (C5). However, the interactions of the variables nutrients and light in aquatic ecosystem are not totally understood and will be the focus of future work in our group.

Table 2 Hydrographic variables determined in the Gulf of Ana María

Station	Depth (m)	Secchi depth (m)	K (m^{-1})	Euphotic depth (m)	EI (SRP)	Trophic status
98	17	3.75	0.53	8.68	2.48	Oligotrophic
101	21	3.75	0.53	8.68	5.33	**Eutrophic**
17	6	5.5	0.36	12.77	2.39	Oligotrophic
26	9	4.25	0.47	9.78	2.31	Oligotrophic
27	9	3.5	0.57	8.07	2.07	Oligotrophic
3	12	6.5	0.31	14.84	2.15	Oligotrophic
170	18	4.5	0.44	10.45	2.64	Oligotrophic
137	5	2.75	0.73	6.30	2.32	Oligotrophic
51	6	3.5	0.57	8.07	2.14	Oligotrophic

3.2 Aquatic Primary Habitability Index

Figures 4 and 5 show the photosynthetic potential of phytoplankton assemblages in the waters of the Gulf of Ana María. Given that stations 26, 27, 51, 98, and 101 showed the same optical water type, their photosynthetic potential curves overlap and similarly occur for the stations 17 and 170. It can be observed that the maximum photosynthetic potential at station 137 is reached shallowest as compared with other stations. This result can be related to the fact that station 137 has the darkest optical water type (C7), which leads to a strong attenuation of UV radiation and consequently reduces the deleterious effects of this waveband. In contrast, the maximum photosynthetic potential at station 3 is reached deepest in the water column since in this station we found the clearest optical water type (C1).

Photosynthetic potential appears to have been affected by UV radiation at the ocean surface, which is indicated by the observed photoinhibition ($P/Ps \approx 0$). Because UV radiation is strongly attenuated in the first meters of the waters column, it is logic to obtain low values of the photosynthetic potential near the surface. Then, from 6 m depth approximately (except station 137) to the bottom of the water column, the photosynthetic potential showed the maximum value in all cases analyzed. Thus, the highest variability of the photosynthetic potential is observed in the first meters of the water column which is determined by the UV radiation fluxes, while the high photosynthetic activity at the bottom of the waters column is maintained due to the high PAR intensity reaching these depths. On the other hand, we did not find differences of the photosynthetic potential between the most and less efficient species using PAR (i.e., $E_S = 2$ W m^{-2} and $E_S = 100$ W m^{-2}), while a small difference was only observed at different times of the days (at 1:00 PM and 4:00 PM). In general, photosynthetic curves at 1:00 PM showed the maximum values in shallower water than the ones at 4:00 PM. In general, the obtained results indicate that there is no limitation by solar radiation to perform the photosynthetic process in the waters of the Gulf of Ana María.

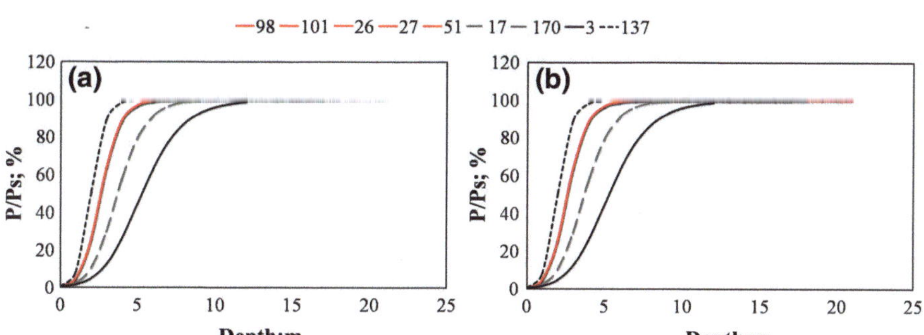

Fig. 4 Photosynthetic potential at 1:00 PM (35°) for **a** most and **b** less efficient species using PAR. Stations with the same color indicate that their photosynthetic curves overlap

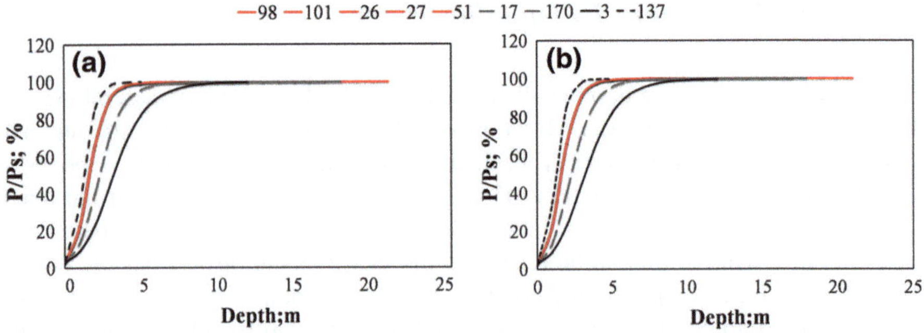

Fig. 5 Photosynthetic potential at 4:00 PM (76°) for **a** most and **b** less efficient species using PAR. Stations with the same color indicate that their photosynthetic curves overlap

Given there were no differences between the photosynthetic potential between the phytoplankton species (i.e., $E_S = 2$ W m^{-2} and $E_S = 100$ W m^{-2}), we did not find considerable differences either in the habitability maps of those species. Thus, in Fig. 6, we only show the spatial distribution of the *APHI* values for the species $E_S = 2$ W m^{-2}, which largely agree with the one for species $E_S = 100$ W m^{-2}. *APHI* values were normalized with respect to its maximum value at station 101. In general, the potential to support the primary productivity of phytoplankton is similar for the times of the day analyzed, being highest in the southwestern gulf and lowest in the central basin. The *APHI* values indicate, in average, an intermediate potential for primary production of phytoplankton ($\sim$50%). The differences of the *APHI* values between stations were determined by the differences in nutrient concentration, which suggest that the latter govern the potential of the waters of the Gulf of Ana María to support phytoplankton productivity.

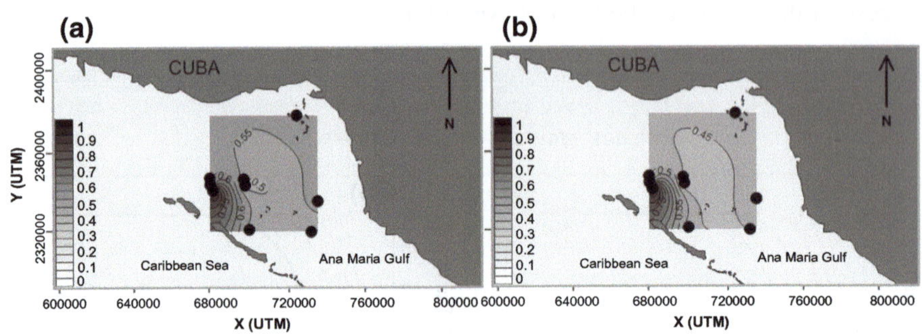

Fig. 6 Spatial distribution of the values of the aquatic primary habitability index primary in the Gulf of Ana María at **a** 1:00 PM and **b** 4:00 PM for the most efficient species using PAR ($E_S = 2$ W m^{-2})

4 Discussion

The coastal waters are less transparent than the open ocean waters, which is determined by the high concentration of sediments and colored dissolved organic matter (CDOM) [34]. In general, three major components contribute to light attenuation in aquatic ecosystems, i.e., the water itself, the dissolved yellow substances (CDOM) and the suspended particulate fraction including phytoplankton (chl-a as a measure of phytoplankton abundance) and tripton (inanimate particulate matter) [23]. The simplest approach to determine the contribution of each of these elements on the light attenuation in the marine water column assumes that their effects are additive in nature [35]. In this case, the total light attenuation coefficient, K_t, is partitioned as follows:

$$K_t = K_{SW} + K_{CDOM} + K_{chl} + K_{tripton} \quad (13)$$

where K_{sw}, K_{CDOM}, $K_{tripton}$ are the partial attenuation coefficients for seawater, CDOM, chl-a, and tripton, respectively.

We suggest that tripton (specifically suspended particulate matter (SPM)) might be present in high concentrations in the water column of the Gulf of Ana María. SPM is an important distinction between estuaries and the open ocean, where SPM concentration is low and attenuation coefficient is more strongly correlated with phytoplankton biomass [36]. It has been reported that the benthic habitat of the stations studied is dominated by mud with some patches of seagrass and corals [37]. Besides, during the studied period, it was reported a high number of atmospheric phenomena such as storms and rainfalls [38], which may lead vertical mixing and therefore resuspension of particulate matter present in high concentrations at the seabed.

The relative shallow Secchi depth obtained in our study can be related to above-mentioned ideas. It has been reported that the Secchi depth can be greater than that observed if only plankton cells are responsible for the light extinction [39]. This observation means that particles other than plankton cells may be the most important factors for determining the Secchi depth in many coastal areas. In many coastal environments, light attenuation by suspended sediments confines the euphotic zone to a small fraction of the surface water column [36]. However, in our case, the euphotic depth was—in almost all stations—higher than the depth of the station (Table 2) which indicates that independent of the existing moderate turbidity (waters C5), still high radiation intensity can penetrate until the bottom given to some extent to the shallowness of the analyzed stations.

In shallow coastal waters, the euphotic zone extent regulates the occurrence and depth distribution of seafloor habitats [40] and therefore the distribution and abundance of photosynthetic organisms. In the Gulf of Ana María, it has been reported a high number of macroalgae, corals, and gorgonians inhabiting the seafloor [41, 42]. Corals live in symbiosis with a dinoflagellate microalga (the so-called zooxanthellae), which need solar radiation and nutrients to develop photosynthesis in the same way that macroalgae and phytoplankton cells do. Thus, it could be considered that intensity of solar radiation reaching the benthic habitats of the Gulf of Ana María is suitable to fuel photosynthesis, including the phytoplankton photosynthesis which has shown the maximum values at the bottom of the studied stations (Figs. 4 and 5).

The temporal variability of photosynthetic rates and depth-normalized total photosynthetic potential is mainly determined by the flux of irradiance (PAR + UV) that penetrates the water column at different times of the day as well as the depth of the station. The flux of radiation at 1:00 PM is greater than at 4:00 PM. The solar angle of incidence at 1:00 PM (35°) allows very little reflection of radiation (i.e., high penetration into the water column) at the air–water interface as compared to the angle of incidence at 4:00 PM which is 76°.

Stations with the highest and smallest *APHI* were 101 and 27, respectively. They showed the same optical water type (C5), but different trophic status (Fig. 3 and Table 2). The station 101 had eutrophic waters for phosphorous, and this type of water body is characterized by a high biological activity, introducing high concentrations of nutrients and primary production rate [18]. On the other hand, the station 27 showed oligotrophic conditions, with very low biological activity, low nutrient concentration and primary productivity, and limited biomass [18]. Thus, we can infer that SRP is the environmental variable limiting primary production in the Gulf of Ana María and not light. This result disagrees with earlier findings for other aquatic systems, for which light availability was reported as the critical environmental factor governing estuarine phytoplankton productivity. In estuaries, nutrient availability is generally adequate to support production [43]; hence, light availability is the most important factor controlling biomass-specific productivity, and consequently, primary productivity is strongly correlated with light availability. The reported oligotrophication in the waters of the Gulf of Ana María could be associated with the damming of the rivers that flow into the gulf. However, more studies are needed to corroborate this suggestion.

5 Conclusions

In this study, we have proposed an optical classification of the waters of the Gulf of Ana María as well as a quantitative index to measure its phytoplankton habitability. It was found a dominance of waters with moderated turbidity (C5) and oligotrophic condition for phosphorus (SRP). The latter appears to be the most important variable influencing the primary habitability of phytoplankton. We obtained that the Gulf of Ana María has an intermediate environmental potential ($\sim$50%) to support phytoplankton productivity. However, it should be noticed that this is a preliminary work. Both the theoretical (APHI) and experimental (sampling in time and space) tools should be refined to reach more accurate conclusions. The APHI can be refined by considering the interactions between the variables, while a more extensive data collection should be done to properly evaluate the spatiotemporal evolution of habitability in the Gulf. This will be addressed in the near future.

References

1. Yool A, Fashman MJR (2001) An examination of the 'continental shelf pump' in an open ocean general circulation model. Global Biogeochem Cycles 15(4):831–844
2. Muller-Karger FE, Varela R, Thunell R, Luerssen R, Hu C, Walsh JJ (2005) The importance of continental margins in the global carbon cycle. Geophys Res Lett 32(1):L01602
3. Flores VFJ, Agraz HC, Benitez DP (2007) Ecosistemas acuáticos costeros: importancia, retos y prioridades para su conservación. In: Sánchez O, Herzig M, Peters RE, Márquez-Huitzil R, Zambrano L (eds) Perspectivas sobre conservación de ecosistemas acuáticos en México, pp 147–167
4. Arndt S, Geneviève L, Nathalie G, Pierre R, Christiane L (2011) Nutrient dynamics and phytoplankton development along an estuary–coastal zone continuum: a model study. J Mar Syst 84:49–66
5. Agboola JI, Kudo I (2014) Different contributions of riverine and oceanic nutrient fluxes supporting primary production in Ishikari Bay. Cont Shelf Res 88:140–150
6. Giraud X, Le Quéré C, Da Cunha LC (2008) Importance of coastal nutrient supply for global ocean biogeochemistry. Glob Biogeochem Cycles 22:GB2025. https://doi.org/10.1029/2006gb002717
7. Alcolado PM, García EE, Arellano-Acosta M (eds) Ecosistema Sabana-Camagüey. Estado actual, avances y desafíos en la protección y uso sostenible de la biodiversidad. Proyecto PNUD/GEF Sabana-Camagüey CUB/98/G32; CUB/99/G81. Editorial Academia, Habana (2007)
8. Kondratiova T, Sosa M (1966) Productividad primaria de las aguas cubanas. Estudios Instituto de Oceanología, Academia de Ciencias de Cuba 2, 21–24
9. V National report to the convention on biological diversity, Republic of Cuba, 2014. Instituto de Ecología y Sistemática y Centro Nacional de Biodiversidad. Ministerio de Ciencia, Tecnología y Medio Ambiente. La Habana (2014)
10. Piñeiro R (2006) Influencia del aporte fluvial en la zona marino costera suroc-cidental del Golfo de Batabanó, Cuba. Rev Cuba Investig Pesq 24(1):28–31
11. Baisre JA, Arboleya Z (2006) Going against the flow: the effect of river damming in Cuban fisheries. Fish Res 81:283–292
12. González-De Zayas R, Lestayo González JA, Merino-Ibarra M, Castillo Sandoval FS (2012) Condiciones hidroquímicas recientes de la zona central del golfo de Ana María, Cuba. Rev Investig Mar 32(2):9–14
13. Piñeiro R, Puga R, González-Sansón G (2006) Bases para el manejo inte-grado del recurso langosta (*Panulirus argus*) en la zona costera sur de Pinar del Río. I. Factores ambientales. Rev Investig Mar 27(3):245–251
14. Baisre JA (2006) Assessment of nitro-gen flows into the Cuban landscape. Biogeochemistry 79:91–108
15. Puga R, Piñeiro R, Cobas S, De León ME, Capetillo N, Alzugaray R (2010) La pesquería de la langosta espinosa, conectividad y cambio cli-mático en Cuba. In: Hernández-Zanuy AC, Alcolado PM (eds) La Biodiversidad en ecosistemas marinos y costeros del litoral de Iberoamérica y el cambio climático. Proceedings of the First Workshop of the network CYTED BIO-DIVMAR. La Habana, Cuba, Instituto de Oceanología, pp 112–131
16. Mendez A (2010) Evolution of the global terrestrial habitability during the last century. In: Sixth astrobiology science conference, pp 26–29, Houston, TX, USA

17. Cardenas R, Perez N, Martinez-Frias J, Martin O (2014) On the habitability of aquaplanets. Challenges 5:284–293
18. Ji ZG (2008) Hydrodynamics and water quality: modeling rivers, lakes, and estuaries. Wiley
19. Platt T, Sathyendranath S (1988) Oceanic primary production: estimation by remote sensing at local and regional scales. Science 241:613–1620
20. Betanzos VA, Garcés RY, Delgado MG, Pis RMA (2012) Variación espacio-temporal y grado de eutrofia de sustancias nutrientes en aguas de los Golfos de Ana María y Guacanayabo, Cuba. Rev Cienc Mar Costeras 4:117–130
21. Sosa BM (2006) Las pesquerías de camarón en Cuba. FAO project report
22. Madronich S (2009) Tropospheric ultraviolet and visible (TUV) radiation model 4.6
23. Kirk JTO (2011) Light and photosynthesis in aquatic ecosystems. Cambridge University Press
24. Shifrin KS (1988) Physical optics of ocean water. American Institute of Physics Translation Series. Springer Science & Business Media
25. Jerlov NG (1976) Optical oceanography. Elsevier Publications Company, Amsterdam
26. Lee Z, Weidemann A, Kindle J, Arnone R, Carder KL, Davis C (2007) Euphotic zone depth: its derivation and implication to ocean-color remote sensing. J Geophys Res 112:C03009: 1–C03009:11
27. Strickland JDH, Parsons TR (1972) A practical handbook of seawater analysis. Fisheries Research Board of Canada, Canada
28. Comar CL, Zscheile FP (1941) Spectroscopic analysis of plant extracts for chlorophyll a and b. Am Soc Plant Biol (ASPB) 16(3):651–653
29. Avila-Alonso D, Baetens JM, Cardenas R, De Baets B (2017) Assessing the effects of ultraviolet radiation on the photosynthetic potential in Archean marine environments. Int J Astrobiol 16(3):271–279
30. Cullen JJ, Neale PJ, Lesser MP (1992) Biological weighting function for the inhibition of phytoplankton photosynthesis by ultraviolet radiation. Science 258:646–650
31. Rodríguez L, Cardenas R, Avila-Alonso D (2014) On the photosynthetic potential in the open oceans. Rev Cuba Fís 30(1):15–17
32. Avila-Alonso D, Cardenas R, Rodríguez L, Alvarez-Salgueiro J (2016) Phytoplankton photosynthetic potential in coastal zones around the world. Rev Cuba Fís 33:62–63
33. Karydis M, Ignatiades L, Moschopoulou N (1983) An index associated with nutrient eutrophication in the marine environment. Estuar Coast Shelf Sci 16:339–344
34. Häder DP, Kumar HD, Smith RC, Worrest RC (2007) Effects of solar UV radiation on aquatic ecosystems and interactions with climate change. Photochem Photobiol Sci 6:267–285
35. Vant WN (1990) Causes of light attenuation in nine New Zealand estuaries: estuarine Coastal and Shelf. Science 31:125–137
36. Cloern JE (1987) Turbidity as a control on phytoplankton biomass and productivity in estuaries. Cont Shelf Res 7:1367–1381
37. Ventura-Díaz Y, Rodríguez-Cueto Y (2012) Hábitats del Golfo de Ana María identificados mediante el empleo de procesamiento digital de imágenes. Rev Investig Mar 32(2):1–8
38. Matos F, Batista LM, Sánchez-Hernández A (2012) Condiciones hidrometeorológicas en la porción central del golfo de Ana María, Cuba. Rev Investig Mar 32(2):15–19
39. Wallin M, Håkanson L, Persson J (1992) Load models for nutrients in coastal areas, especially from fish farms (in Swedish with English summary). Nordiska ministerrå det, Copenhagen
40. Luhtala H, Tolvanen H (2013) Optimizing the use of Secchi depth as a proxy for euphotic depth in coastal waters: an empirical study from the Baltic Sea. ISPRS Int J Geo-Inf 2:1153–1168
41. Suárez AM, Martínez-Daranas B, Guimarais-Bermejo M, Volta R (2013) Macroalgas del golfo de Ana María, SE de Cuba. Rev Investig Mar 33(2):1–6

42. González-Díaz P, Perera-Pérez O, Pérez-García JA, Hernández-Fernández L (2012) Biodiversidad de corales, gorgonias y esponjas en un sector del golfo de Ana María. Rev Investig Mar 32(2):20–29
43. Nixon SW (1981) Freshwater inputs and estuarine productivity. In: Cross R, Williams D (eds) National symposium on freshwater inflow to estuaries, U.S.

On the Subaquatic Light Fields in Lakes of Southern Chile and Their Photosynthetic Potential

Lien Rodríguez López, Rolando Cárdenas, Oscar Parra, Roberto Urrutia, Lisdelys González and Rebeca Martínez

Abstract The district of Chileans Araucanian Lakes is a group of lakes of glacial origin that are located between 39° and 42°, Lat. S and 71° and 72° Long. W that are characterized by presenting oligotrophic conditions. The last of the chain of lakes of Valdivia Basin River is the Riñihue, a temperate monomítica lake, reason why its study is of great importance, since it reflects any disturbance in the chain. It is relevant to have precise spectra of the photosynthetically active and ultraviolet radiation of the location area to calculate the photosynthetic potential in these lakes. In this paper, an assessment is made about the applicability of the COART radiative transport model in freshwater aquatic systems.

Keywords Radiative · Phytoplankton · Aquatic

L. R. López (✉) · O. Parra · R. Urrutia · R. Martínez
Environmental Sciences Center (EULA), University of Concepción, Concepción, Chile
e-mail: lienrodriguezlopez@gmail.com

O. Parra
e-mail: oparra@udec.cl

R. Urrutia
e-mail: rurrutia@udec.cl

R. Martínez
e-mail: rmartinezretureta@gmail.com

R. Cárdenas
Planetary Science Laboratory, Universidad Central "Marta Abreu" de Las Villas, Santa Clara, Cuba
e-mail: rcardenas@uclv.edu.cu

L. González
Faculty of Engineering, University of Concepción, Concepción, Chile
e-mail: lisdegonzalez@udec.cl

R. Cárdenas et al. (eds.), *Proceedings of the 2nd International Conference on BioGeoSciences*, https://doi.org/10.1007/978-3-030-04233-2_9

1 Introduction

The continental aquatic ecosystems are characterized by a great complexity in which different biotic and abiotic elements interact in several temporal and spatial scales in very diverse ways [1]. This causes that the study, management actions, and prediction of possible future states are complex tasks and difficult to research using classical scientific methods. One way to reduce the complexity with which one works in environmental management and research is the development and use of physical–mathematical models to apply to the ecology that represent the most important processes and elements of the ecosystem. The modeling of radiative transport in atmosphere and ocean has been much more worked than in continental aquatic systems (freshwater); therefore, the development of scientific research is indispensable in order to evaluate the applicability of ocean radiative transport models for lacustrine ecosystems [2].

In the last three decades, there has been a decrease in ozone concentrations in the southern hemisphere [3]; as a consequence, this hemisphere has received more UV radiation than the northern hemisphere [4]. In terms of its terrestrial distribution, ozone varies over time, with daily or seasonal periods; these variations are caused by the circulation of stratospheric winds and by the production and chemical destruction of ozone. It is generally found in lower concentration in Ecuador and in greater abundance in the poles. Due to the special climatic and meteorological conditions in Antarctica and nowhere else on the planet, a hole is forming in the ozone layer. The difference between the ozone content in the Polar Regions comes from different meteorological patterns [3].

Chile, due to its proximity to Antarctica, has extreme radiation conditions, so it is a unique experimental natural scenario to prove what this increase would mean for ecosystems and aquatic organisms, in addition to the high levels of radiation it receives, due to the possible implications and uncertainties that climate change would generate in this country. Due to the physical–geographical characteristics of this territory, the highest percentage of the chilean territory presents great elevations, according to studies carried out in the northern zone [5], over 1000 msnm, UV radiation increases by 10%, so the same. It can vary with altitude and latitude [6, 7]. In relation to the Chilean territory from Valparaiso to the lakes, the sun's rays are classified as dangerous, specifically between 11:00 and 16:00 h according to Conac studies [8].

On the other hand, Chile has several lake districts throughout the country; the Araucanos lake district located in the southern part of the American continent is considered one of the most important. According to the ecosystem value, this water is for different uses: represent a reservoir the drinking water, for irrigation, agriculture and represent a high landscape, cultural and tourist value. Therefore, the investigation of these lake ecosystems has aroused an interest and relevance in recent years. The present investigation will contribute with the scientific bases obtained from one of these lakes: the Riñihue Lake.

According to the ecosystem value, is water is for different uses: represent a reservoir the drinking wáter, irrigation, agriculture.

2 Materials and Methods

2.1 Description of the Study Area

Lake Riñihue is located at 39°50′ south latitude and 72°20′ west longitude. It is a pre-Cordilleran lake in the province of Valdivia in southern Chile, it is part of a chain of six lakes (Araucanian Lakes). It has an altitude of 117 m above sea level, with a basin of 4500 km^2 of which 77.5 km^2 correspond to the lake. The Riñihue presents a glacial origin. The waters of this lake discharge into a river that receives different names before joining the sea. From its beginning to the encounter with the Collileufu River it is called San Pedro, from this point until the junction with the Cruces River it is called Calle-Calle and from there to its drainage in the Corral Bay it is called the Valdivia River. The river has a course of 110 km until its mouth [9]. The climate of the area is temperate of rainy type; the superficial temperatures of the water vary of 9–20 °C during the year. There is thermal stratification (>2–3 °C) for 9–10 months of the year, beginning in September [10–12]. This lake is monomictic warm, which means that during the summer period it is stratified and, during the winter months, the water column is mixed. The thermal amplitude rises inwards and is moderate due to the influence of the sea.

As hydrological characteristics, the precipitations take place throughout the year and increase from north to south, border the 1167 mm per year, has 27 km maximum length and 5 km wide, a volume of 12.6 km^3 with a maximum depth of 323 m. The level of the lake varies approximately 2 m, registering the highest values in winter and the lowest in summer; the variations are closely related to the meteorological events of rain (Fig. 1).

2.2 Obtaining of the Data Series

In the study, a time series was taken from 1987 to 2016 (30 years). Experimental data reported in the literature that were taken from several studies: studies conducted by Professor Hugo Campos in Limnological study of Araucanian Lakes (Chile) in 1984, the study: Partial operation of the minimum control network of Riñihue Lakes and Ranco UACH-DGA, 1988. Study: Redefinition of the minimum network of lakes POCH AMBIENTAL SA, 2009. Study: Final report No. 1. Phytoplankton analysis in water samples. Tender No. 1019-98-LE13. Study: Phytoplankton analysis in water samples, Eula, 2012. Study: Analysis and reformulation network monitoring lakes region of the rivers, DGA 2016.

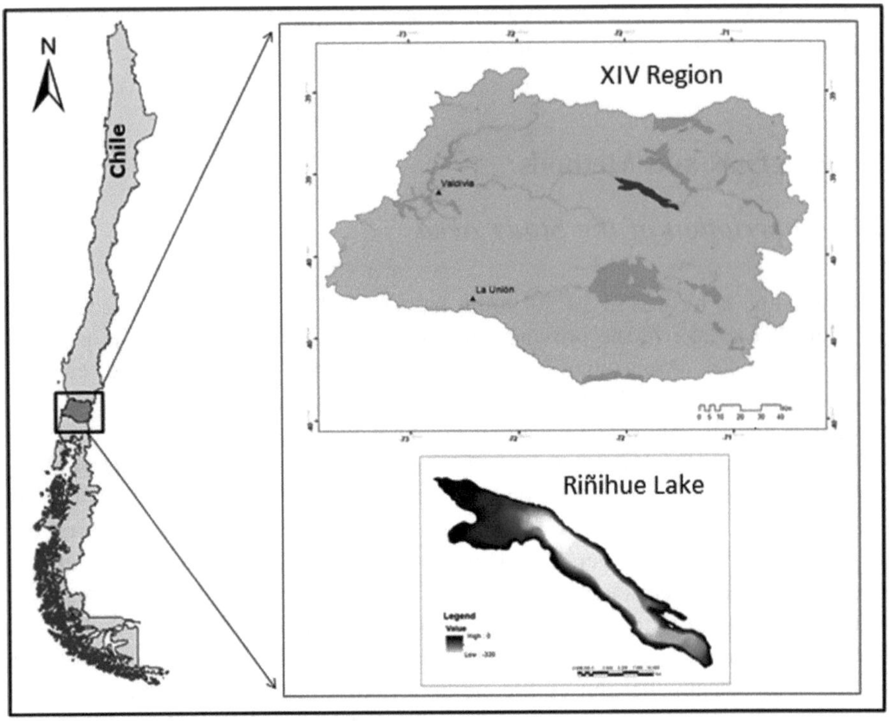

Fig. 1 Geographical location of the study area

2.3 Radiative Transport Model (COART)

The COART radiative transport model (Coupled Ocean Atmosphere Radiative Transfer) is a model of unified radiative transport developed by NASA Solar and Cloud Radiation Group, in the sense that it considers the ocean as the lower layer of the atmosphere. It also includes the influence of turbulence on the penetration of light in the upper oceanic layers. It has been implemented for saltwater, the potential for its application in freshwater was tested in a previous work [2] and with this paper, and a continuity will be given for its future application in continental aquatic ecosystems.

2.4 Evolution of Phytoplankton Species Through Historical Records

The phytoplankton data were taken in three years within the study series; the first is in 1987 where Professor Hugo Campos carried out a monitoring campaign on Lake

Riñihue and describes the 43 phytoplankton groups found, and the second period corresponds to a study of the Center for Environmental Studies (EULA) conducted by Professor Oscar Parra and Dalba Avilés in 1997 where the specific wealth increased to 68 taxa, and the last interval was in 2015, data that was taken from the monitoring campaign carried out by the EULA according to the disposition of the General Direction of Waters (DGA) of Chile, this time increased the wealth being of 107 groups.

3 Results and Discussion

3.1 Relationship Between UV and Chlorophyll a

The following graphs show how the UV values (on the surface and at 10 m depth) obtained from modeling with COART and chlorophyll a values during the study time series are related. The graphs were constructed for several zenital angles representative of Central Chile from 30° to 60° with an interval of 10. The same for the autumn season subscribes below:

Autumn

See Figs. 2, 3, 4, and 5.

During this season, it is observed that incident UV radiation at the surface level has a high value in the study period, even though it is attenuated in the water column it shows high values registered at 10 m depth; UV radiation is cumulative in lake sediments and over time it could mean extreme conditions for aquatic organisms in case of disturbance. Due to the extreme conditions, UV radiation in the 10 m depth remain stable in the period studied and may have caused an adaptation to aquatic species to these values of radiation.

Autumn

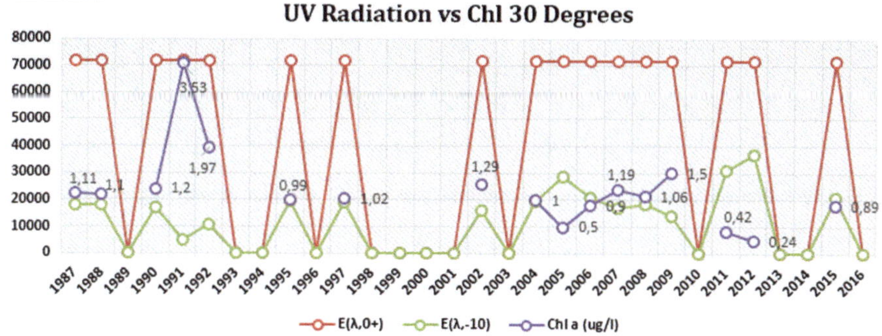

Fig. 2 UV radiation versus Chl a in the Autumn season, 30° of solar zenital angle

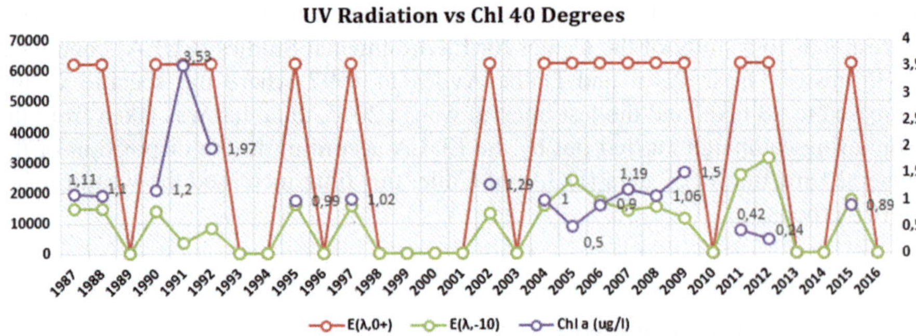

Fig. 3 UV radiation versus Chl a in the Autumn season, 40° of solar zenital angle

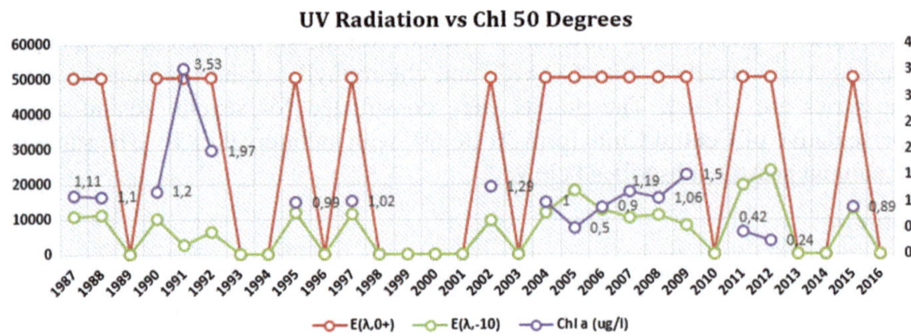

Fig. 4 UV radiation versus Chl a in the Autumn season, 50° of solar zenital angle

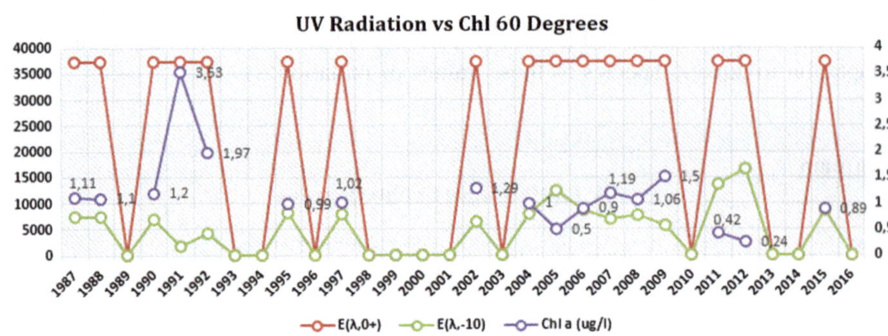

Fig. 5 UV radiation versus Chl a in the Autumn season, 60° of solar zenital angle

Winter

See Figs. 6, 7, 8, and 9.

In this season, chlorophyll a has a greater variability due to the instability in the water column of the phytoplankton suspension due to the existing mixture,

Winter

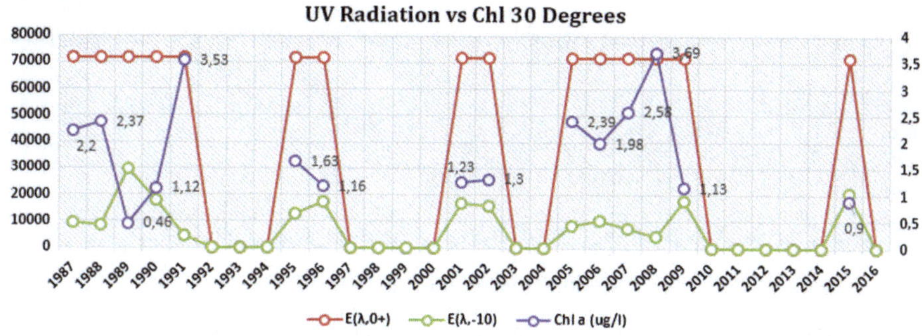

Fig. 6 UV radiation vs Chl a in the Winter season, 30° of solar zenital angle

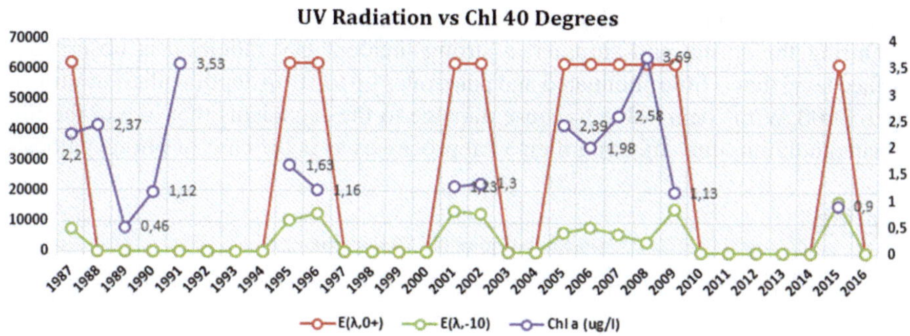

Fig. 7 UV radiation versus Chl a in the Winter season, 40° of solar zenital angle

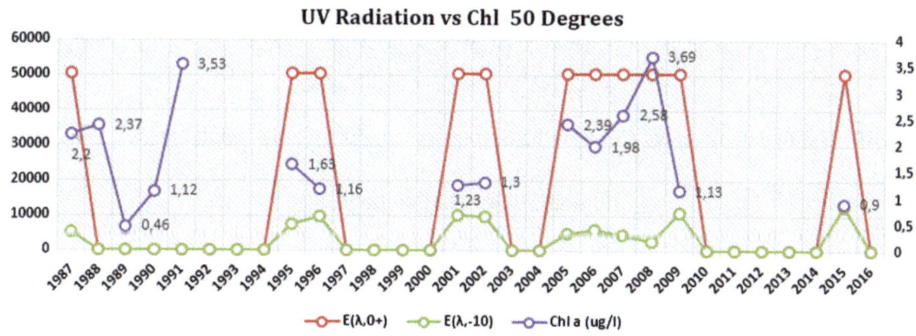

Fig. 8 UV radiation versus Chl a in the Winter season, 50° of solar zenital angle

the average radiation throughout the year does not present significant variations at the surface level, while the UV that penetrates in the 10 m depth it is much smaller than in the other stations characteristic that can be derived from the mixture itself and function as a barrier to the passage of light.

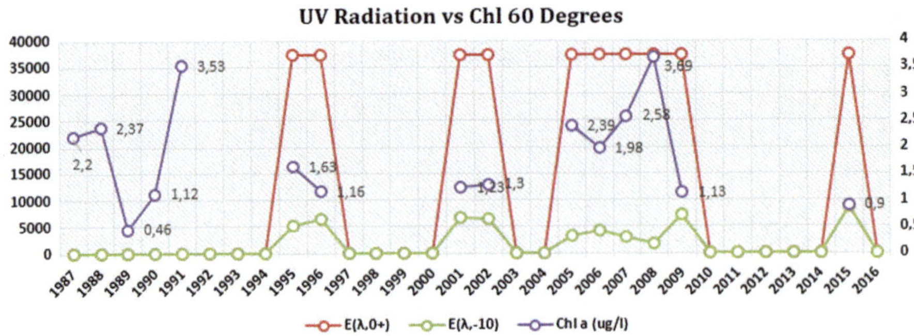

Fig. 9 UV radiation versus Chl a in the Winter season, 60° of solar zenital angle

Spring

See Figs. 10, 11, 12, and 13.

During the spring and summer, a strong incident UV radiation is observed at surface level that is little attenuated and penetrates up to 10 m depth with a radiation of up to 3000 W/m^2 being able to cause damage to the organisms that live in the illuminated column, thus affecting their vital processes as is the case of photosynthesis.

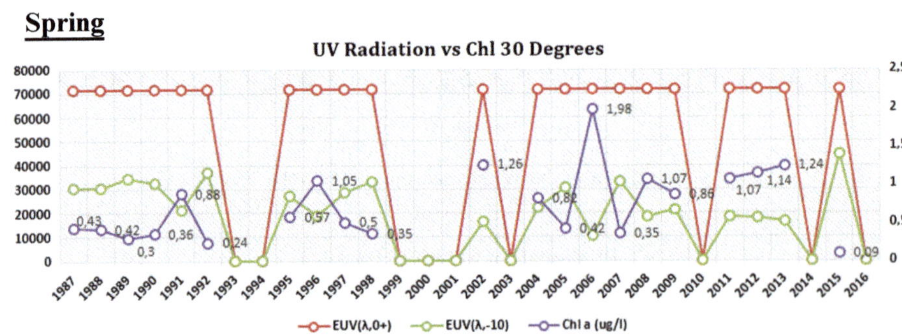

Fig. 10 UV radiation versus Chl a in the Spring season, 30° of solar zenital angle

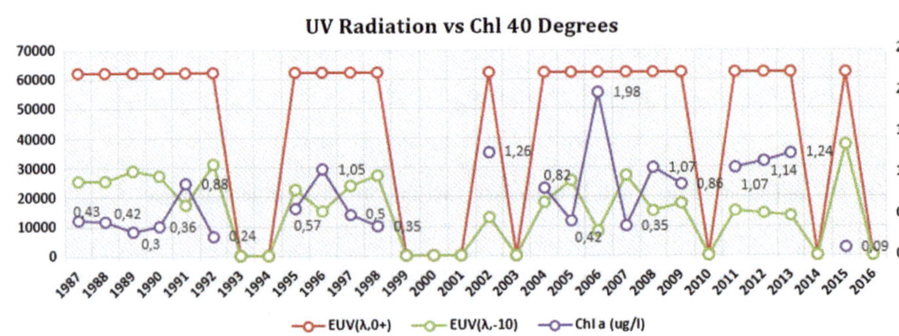

Fig. 11 UV radiation versus Chl a in the Spring season, 40° of solar zenital angle

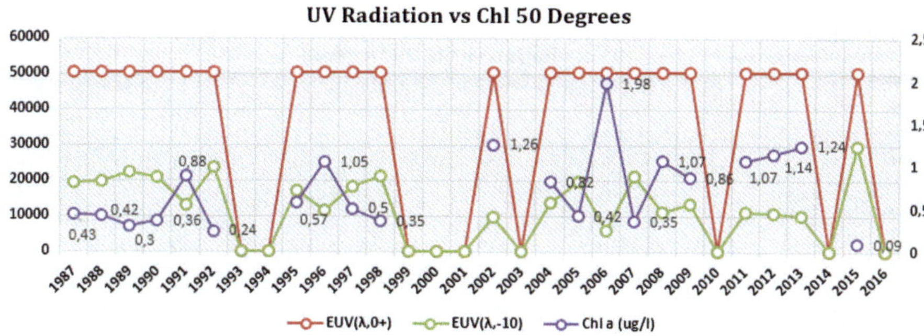

Fig. 12 UV radiation versus Chl a in the Spring season, 50° of solar zenital angle

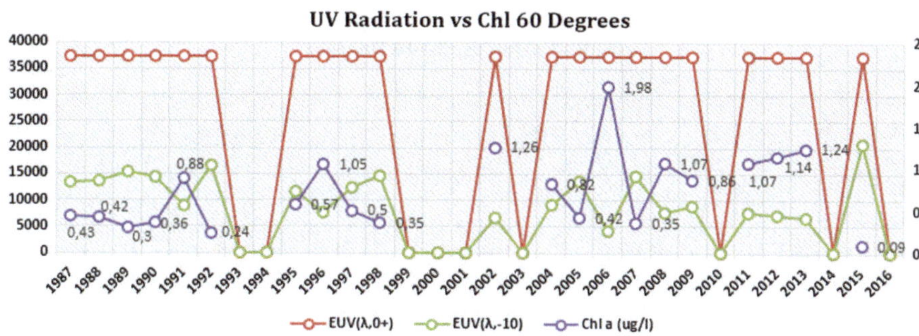

Fig. 13 UV radiation versus Chl a in the Spring season, 60° of solar zenital angle

Summer

See Figs. 14, 15, 16, and 17.

During the whole period studied (temporary series 30 years), the incident UV radiation had a simulated behavior at the surface level, while the UV radiation calculated for the 10 m depth had significant seasonal differences being higher in the spring and summer seasons by the penetration and light stability in the water column which may represent damages to the primary producers regarding the inhibition of the photosynthetic process. It was observed that while increasing the zenital angle the incidence of UV radiation decreases by approximately 100 W/m^2 every 10°. In Autumn of 1991, there was a maximum of incident UV radiation at the surface level coinciding with the increase in the chlorophyll a value reported for this season in the same year (1991), in a previous work [13] a decrease of ozone in Autumn that could cause this increase in UV. The same happens for the spring of 2006, where an increase in the chlorophyll value coincides with an increase in the UV radiation incident on the surface of the lake.

Using the IBM SPSS 22 statistical software, a Pearson's correlation was performed for UV variables at 10 m depth and for chlorophyll a. Table 1 shows these

Summer

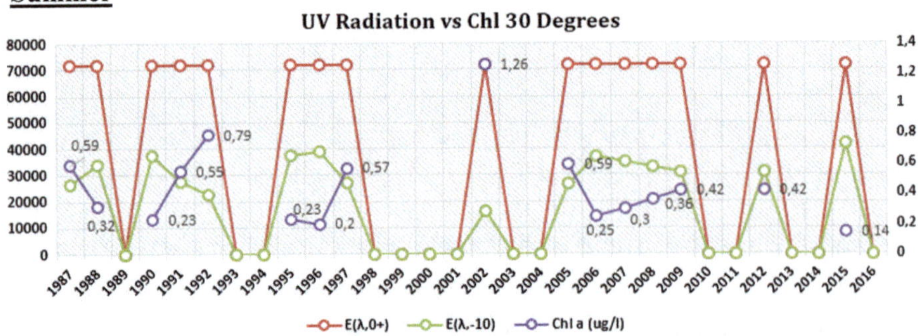

Fig. 14 UV radiation versus Chl a in the Summer season, 30° of solar zenital angle

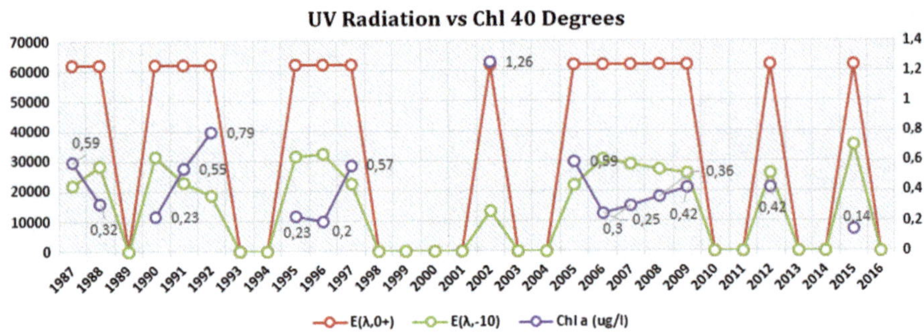

Fig. 15 UV radiation versus Chl a in the Summer season, 40° of solar zenital angle

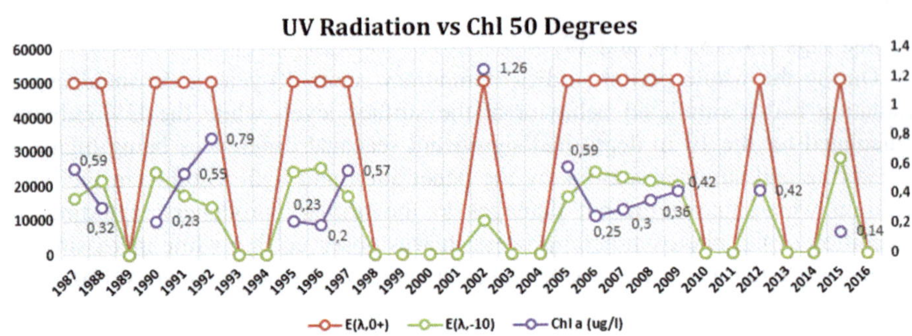

Fig. 16 UV radiation versus Chl a in the Summer season, 50° of solar zenital angle

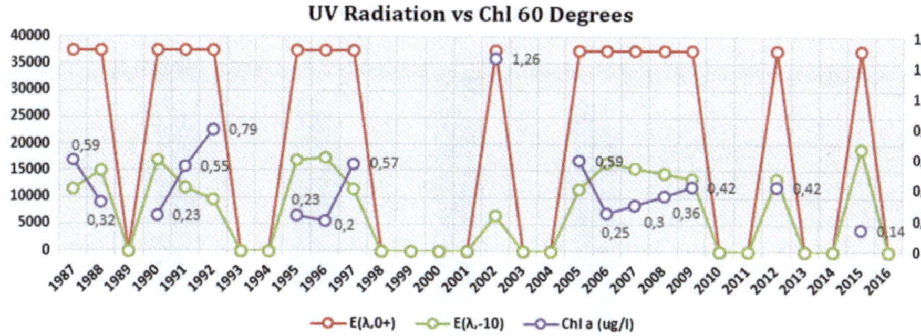

Fig. 17 UV radiation versus Chl a in the Summer season, 60° of solar zenital angle

Table 1 Pearson's correlation for UV variables 10 m and Chl a (30° and 60°) for each season of the year

Seasons	30°	60°
Autumn	−0.877	−0.853
Winter	−0.932	−0.952
Spring	−0.954	−0.943
Summer	−0.969	−0.96

correlations for each season of the year and for the extreme zenital solar arrays that were taken into consideration (30° and 60°).

Highly negative correlation between these variables was obtained for all stations, in all cases greater than −0.86, which could mean that the UV variable is the dominant environmental variable and that it may be limiting the growth of the phytoplanktonic groups in the months where it reaches its maximum values.

3.2 *Optical Classification of the Lake*

Transparency values (meters) were recorded, measured by Secchi disk in the aforementioned monitoring campaigns in order to know the behavior of the passage of light through the water column and later used to calculate the attenuation coefficient of the light (PAR); the values obtained in the time series for each year were averaged by season of the year. Figure 18 shows the values of the transparency behavior of the water column during time series and seasonal differences.

The lower values are reflected in the Autumn season due to possible mixtures in the water column, which could cause the turbidity to prevent the passage of light in it. The greatest transparency is in summer, because the characteristics of the station itself have greater incident radiation, and in this period the lake is stratified, so the circulation of the water body is not much and the equilibrium allows a greater passage of water the light.

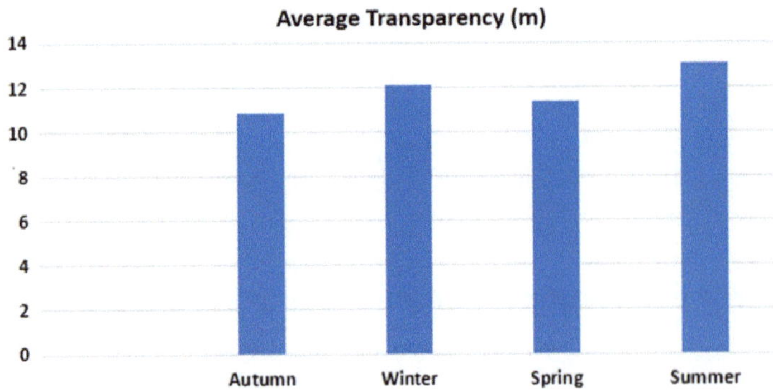

Fig. 18 Average transparency during the time series by season of the year

Table 2 Light attenuation coefficient by season in the time series

Kd (1987–2016)				
Radiation	Winter	Spring	Summer	Autumn
UV	0.95	0.87	0.86	1.09
PAR	0.16	0.16	0.15	0.18

The attenuation coefficient of the light PAR (Kd) was calculated for each season of the year (Table 2) by formula (1):

$$\mathrm{Kd} = 2/Z(\mathrm{Secchi}) \tag{1}$$

where Kd is (the attenuation coefficient) and *Z*(Secchi) is the transparency average the Secchi disk.

The attenuation coefficient corresponding to UV light was calculated through the Lambert Beer law of Optics, evaluated at depth $z = 10$ m:

$$E(\lambda, 10) = E(\lambda, 0-)\exp[-\mathrm{Kd}(\lambda)10] \tag{2}$$

The attenuation coefficients for PAR and UV radiation are shown in Table 2.

According to the attenuation coefficients of the PAR and UV radiation obtained, the Riñihue Lake for photosynthetic active radiation behaves similarly to the Jerlov type I oceanic waters (Fig. 19, Source: [14]) and for the attenuation coefficients corresponding to UV light, it is an intermediate between the oceanic waters type III and the coastal ones C1 [15]. In all cases, the optical classification of this lake is in correspondence with the trophic characteristics of an oligotrophic aquatic system (very clear waters and with low productivity).

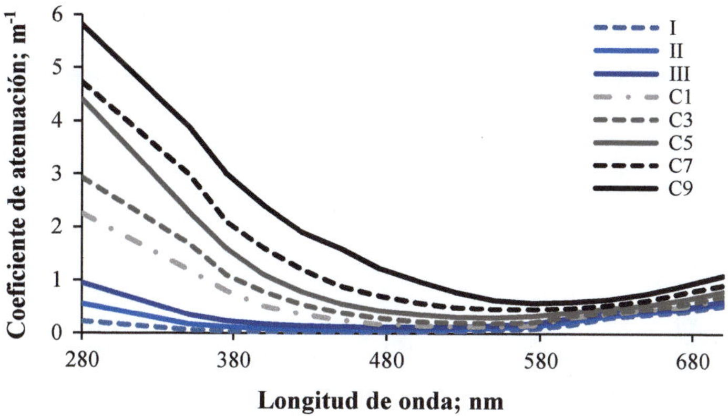

Fig. 19 Jerlov classification of coastal and oceanic waters

3.3 Results of Evolution of Phytoplankton Species Through Historical Records

To know the evolution in these 30 years of the phytoplanktonic groups of this lake, three periods of study were taken into account: The first time period considered was in 1987 through the results obtained in the monitoring campaign of Professor Hugo Campos, the second period considered corresponds to the year 1997 where a campaign was conducted by the Center for Environmental Studies (EULA) led by Professor Oscar Parra and Dalba Avilés, and the last stage of analysis was the one corresponding to 2015 in the monitoring campaign as provided by the Directorate Water General of Chile (DGA).

In 1987, 43 phytoplankton species were described, of which: belong to 20 Bacillariophyceae (being the most abundant group), 16 Chlorophyceae, 2 Chrysophyceae, 3 Cyanophyceae, and 2 Dinophyceae, in the same period a minimum cell density of 268.287 cel/L was found and maximum density of 1.627285 (cel/L); the species with the highest density corresponded to the types: *Asterionella formosa* and *Melosira husteidii* (both of the Bacillariophyceae group). The minimum biomass was 195.6 g/m^2, and the maximum biomass was 4192.6 g/m^2, respectively. The average chlorophyll a value during the campaign was 0.9 ug/l.

After a decade in the campaign of 1997, 68 phytoplankton species were reported: 35 species of Bacillariophyceae, 19 Chlorophyceae, 3 Chrysophyceae, 5 Cryptophyceae, 3 Cyanophyceae, and 2 Dinophyceae; the number of species of the most abundant group (Bacillariophyceae) was increased, a maximum cell density of 1. 700238 cel/L and a minimum of 14.188 cel/L was reported, which increased considerably in this last decade, something that was to be expected considering the increase of the groups, the higher density species that were described were: *Rhodomonas lacustris, Anabaena spiroides, Melosira distans, Asterionella formosa,* and *Sphaerocystis schroeteri*. Among them, a cyanobacteria (*A. spiroides*) characterized by being toxic and form blooms in systems of higher degree of trophy.

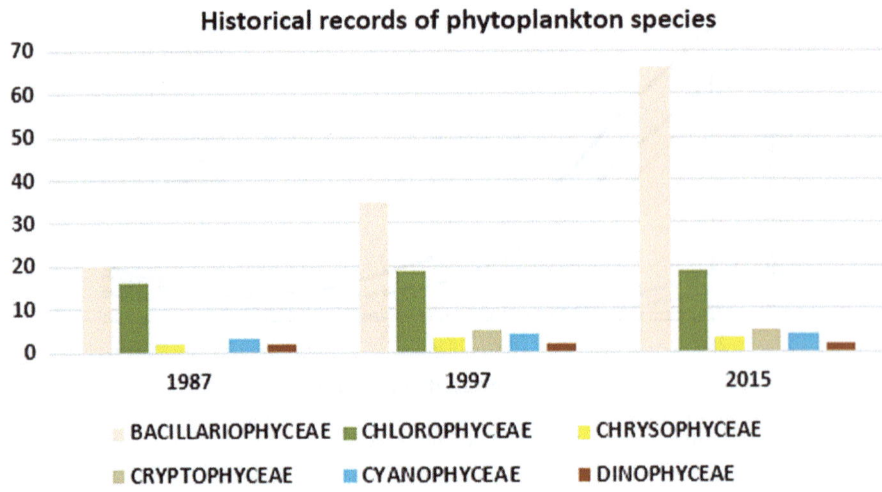

Fig. 20 Historical records of phytoplankton species

In this period, the Dinophyceae *Ceratium hirundinella*, a species reported as invasive in Chile for having a high degree of expansion and invading most of the continental systems of the south of this country, was detected for the first time in this freshwater system. The average value of Chl a during this campaign was 1.02 ug/l.

The last stage of analysis corresponding to the year 2015 where it is described that this lake has a specific richness of 107 species, of them: 66 Bacillariophyceae, 19 Chlorophyceae, 3 Chrysophyceae, 5 Cryptophyceae, 4 Cyanophyceae, and 2 Dinophyceae, increasing, as well that in the previous record, the Bacillariophyceae group, a maximum cell density of 207.927 cel/L was reported. The presence of the invasive species is once again found: *C. hirundinella* but of greater concern, the infestation species *Didymosphenia geminata*, which has invaded a large number of freshwater aquatic ecosystems and has been associated with flood conditions [16], is reported in its tributary (Rio Enco) systems characterized by low nutrients and studies suggest that it has been related to supporting high values of radiation. Among the most abundant species in this year are: *Tabellaria fenestatra, Aulacoseira granulata,* and *Fragilaria crotonensis*, continuing the Diatoms being the dominant species in this lake.

The graph (Fig. 20) shows the evolution of these groups in the series of time studied according to historical records.

4 Conclusions

- The COART radiative transfer model was applied again to a freshwater ecosystem, showing as a result the behavior of UV radiation at the surface level and at 10 m depth.

- Graphs of UV and chlorophyll a radiation were obtained for each season of the year in different zenital solar angles from 30° to 60° with separation of 10, it could be determined that as the angle increases every 10° decreases with a rate 100 W/m^2 the value of the radiation.
- The statistical software IBM SPSS Statistics was used to perform Pearson's correlation between the UV variables at 10 m depth and the chlorophyll variable during all seasons of the year; this analysis shows a highly negative correlation between the study variables with higher values in all the cases at −0.86, the highest correlation was found in the summer season, which can be means that UV is the dominant environmental variable and that it may be limiting the growth of these phytoplankton groups during the seasons of the year when the UV takes its maximum values.
- An optical classification of the lake is established in the series of time studied (30 years) where the attenuation coefficient (kd) of the light was calculated for each season of the year (Table 1) which was corresponding to the type of water C1 according to the Jerlov classification of the oceanic and coastal waters.
- An analysis of the phytoplankton groups was carried out in three years of the time series studied to see their evolution, in 1987 where information was recorded in the monitoring campaign carried out by Professor Hugo Campos, the second year of study was in 1997 during another campaign of the center of environmental studies (EULA) directed by Professor Oscar Parra and Dalba Aviles and the last year 2015 campaign according to the disposition of the General Directorate of Water of Chile DGA. There is an evolution in the most abundant group: Bacillariophyceae with a wealth in 1987 of 20 groups and in 2015 they increased to 66.
- In 1997, the dinophyceae *C. hirundinella* was found for the first time in this aquatic ecosystem, a species reported as invasive in Chile for having a high degree of expansion and invading most of the continental systems of the south of this country.
- In 2015, the presence of the invasive species is once again found: *C. hirundinella* but of greater concern, the infestation species *D. geminata*, which has invaded a large number of freshwater aquatic ecosystems and has been associated with flood conditions, is reported in its tributary (Rio Enco). Systems characterized by low nutrients and studies suggest that it has been related to supporting high values of radiation

References

1. Pearson RG, Dawson TP (2003) Predicting the impacts of climate change on the distribution of species: are bioclimate envelope models useful? Glob Ecol Biogeogr 12:361–371. https://doi.org/10.1046/j.1466-822x.2003.00042.x
2. Rodríguez-lópez L, Cárdenas R, Parra O, Urrutia R, González L, Martínez R, De Concepción U (2017) II CONFERENCIA INTERNACIONAL DE BIOGEOCIENCIAS Sobre la aplicabilidad del modelo de transporte radiativo COART a sistemas acuáticos continentales On the applicability of radiative transfer model COART to continental aquatic systems, 1–9

3. Agustí S (2007) Aumento de la radiación ultravioleta: Impacto sobre los océanos polares, Impacto Del Calentamiento Glob. Sobre Los Ecosistemas Polares 25–46
4. Agustí S, Llabrés M, Carreja B, Fernández M, Duarte CM (2015) Contrasting sensitivity of marine biota to UV-B radiation between southern and northern hemispheres. Estuaries Coasts. 38:1126–1133. https://doi.org/10.1007/s12237-014-9790-9
5. Rivas M, Rojas E, Madronich S (2008) Aumento del Índice Solar Ultravioleta con la Altura Solar Ultraviolet Index Increase With Altitude, Ingeniare. Rev Chil Ing 16:383–388. https://doi.org/10.4067/s0718-33052008000200013
6. Cornejo L, Martín-Pomares L, Alarcon D, Blanco J, Polo J (2017) A through analysis of solar irradiation measurements in the region of Arica Parinacota, Chile. Renew Energy 112:197–208. https://doi.org/10.1016/j.renene.2017.04.012
7. Molina A, Falvey M, Rondanelli R (2017) A solar radiation database for Chile. Sci Rep 7:1–11. https://doi.org/10.1038/s41598-017-13761-x
8. Conac, INFORME DE RADIACIÓN UV Y CÁNCER DE PIEL EN CHILE 2016, Santiago de Chile, 2016. http://ambiente.usach.cl/uv
9. Campos H, Bucarey Bahamondes elys, Arenas jn (1974) Estudios Limnológicos del lago Rinihue y rio Valdivia (Chile).
10. Parra O y Hugo Campos LZ (1987) Wladimir Steffen, Gloria Aguero, Limnology of Lake Riñihue, 239–357
11. Woelfl S, Villalobos L, Parra Ó (2003) Trophic parameters and method validation in Lake Riñihue (North Patagonia: Chile) from 1978, through 1997. Rev Chil Hist Nat 76:459–474. https://doi.org/10.4067/S0716-078X2003000300010
12. Imberger J, Javam A, Campos H, Hamilton DP, Villalobos L (2001) A modelling assessment of potential for eutrophication of Riñihue Lake, CHILE, 101–125
13. Rodríguez L, Parra O, Cárdenas R, Urrutia R, González L, Martínez R (2017) On the applicability of radiative transfer model COART to continental aquatic systems. Editorial Samuel Feijoo (Indexed Scielo)
14. Rodríguez L, Cárdenas R, Ávila-Alonso D (2014) On the photosynthetic potential in the open oceans. Rev Cuba Fis 31:15–17
15. Avila-Alonso D, Cardenas R, Rodriguez L, Alvarez-Salgueiro J (2016) Phytoplankton photosynthetic potential in coastal zones around the world. Rev Cuba Fis 33:62–64
16. Beamud SG, Baffico G, Reid B, Torres R, Gonzalez-Polo M, Pedrozo F, Diaz M (2016) Photosynthetic performance associated with phosphorus availability in mats of *Didymosphenia geminata* (Bacillariophyceae) from Patagonia (Argentina and Chile). Phycologia 55:118–125. https://doi.org/10.2216/15-83.1

Coarse Detrital Deposits from Hurricane Wilma on the Western Coast of Cojimar, Havana, Cuba

Reinaldo Rojas-Consuegra, Jorge Isaac-Mengana, Felipe Matos Pupo and Matthew Charles Peros

Abstract Hurricane Wilma produced huge waves that hit the western coast of Cuba. The coast sector located west of the town of Cojimar remained under extremely heavy wave action and unidirectional currents for almost 72 h. Sediment tests provided new knowledge on the sediment processes that took place in the sea–land interphase. The effects of this anomalous event were evident in elements such as detritus, coastal boulders and vegetation, the soil and existing anthropogenic materials. Results make evident important regularities of high methodological value. It is highlighted in this work that the elimination of the coastal vegetation increases the negative effects of storms, increasing vulnerability in the coastal zone in the face of these extreme events.

Keywords Coastal boulders · Detritus · Hurricane Wilma · Sediment processes Vulnerability

1 Introduction

The season of 2005 year turned out to be exceptional from many points of view. It was a "record" season in terms of number of tropical storms and hurricanes, moving the season of 1933 to a second place in the Atlantic area. In the latter, 21 cyclonic organisms were formed; while in 2005, there were 27 of them. It was also "record" in the number of hurricanes, registering 15, with 7 of them of great intensity, surpassed only by the season of 1950. In addition, it has been the longest season,

R. Rojas-Consuegra · J. Isaac-Mengana
Museo Nacional de Historia Natural de Cuba, Havana, Cuba

F. M. Pupo (✉)
Centro de Investigaciones de Ecosistemas Costeros, Cayo Coco, Cuba
e-mail: fmatos@ciec.cu

M. C. Peros
Department of Environmental Studies and Geography, Bishop's University, Sherbrooke, Québec, Canada

R. Cárdenas et al. (eds.), *Proceedings of the 2nd International Conference on BioGeoSciences*, https://doi.org/10.1007/978-3-030-04233-2_10

since it finalized on January 6, 2006, with the loss of the cyclonic characteristics of the last organism of the season ("Zeta"). The 1954 season was also very extensive but ended on January 5, 1955 [1].

During 2005, our country was hit directly or indirectly by five tropical cyclones, directly by Tropical Storm Arlene in June and by Hurricane Dennis (Category 4) in July, being the first intense hurricane to affect the country in a month of July, while hurricanes Katrina, Rita, and Wilma hit several regions of the country with strong winds, heavy rains, and coastal flooding, although their centers did not pass through the national territory.

The 2005 season placed three hurricanes among the first six places of lowest central pressure in the Atlantic basin. Wilma (882 hPa) displaced Hurricane Gilbert (888 hPa), which since 1988 occupied the top spot, given the intensity and the experienced trajectory (Fig. 1), in October 2005, Hurricane Wilma produced huge waves that hit the western coast of Cuba.

The coastal floods generated in Havana City were very severe and prolonged. The coast sector located west of the town of Cojimar remained under extremely heavy wave action and unidirectional currents (NW to SE) for almost 72 h. This event brought about unusual effects on the intertidal and supratidal area, which are described in this work (Picture 1). According to Fig. 1, the points of the trajectory closest to Cojimar correspond to October 23.

Fig. 1 Trajectory of Hurricane Wilma (Category 5). Hurricane season of 2005

Picture 1 Big waves (3–4 m) impacted the town of Cojimar during almost 72 h. *Courtesy* of Emilio Alfaro

Documenting the consequences of unusual severe events is a way to understand natural phenomena that affect the environment and to improve our capacity to face them, mitigating the damages to economic infrastructure and the environment. However, referring to extreme wave events, it is important to consider that the number of intense hurricanes in the different ocean basins is increasing. In general, changes are being made in terms of frequency and intensity of these meteorological phenomena [2].

The study of boulders moved by hurricanes in present times is an excellent way to use as analogues to know the magnitude of hurricanes in the past. Using other geological files (analysis of sedimentological samples deposited in coastal lagoons), new records of paleohurricanes have been identified in Cuba [3]. The complement of both proxies' techniques is basic to improve the results of this type of research.

2 Materials and Methods

The trajectory of Hurricane Wilma was obtained by running the Hurricane Watch'99 software.

Using methods and techniques to gather data from sediment analysis [4, 5], new information on the interaction taking place on the land–sea interphase has been gathered.

In this case, land sediments induced by wave action in unidirectional flows (wash over fans) were assessed [6]. This is a distinctive feature of this kind of deposit as it is really unusual. Regularly, it is the other way around, i.e., from land to sea, transported by river currents or other seasonal flows.

The sources of detritus (any material, fragmented or not found in this kind of deposit) are mainly coastal deposits. They consist of ancient accumulations of clasts or coastal fragments, biogenic materials temporarily accumulated in the intertidal or supratidal adjacent area, rock fragments from the intertidal zone, garbage and construction waste and debris from the seabed during the hurricane as waves turn deeper and consequently remove underwater detritus.

Based on these criteria, the distribution of the different materials was measured (using 5 and 50 m measuring tapes). Behavior patterns during flow and accumulation, direction of the main flow (using a geological compass), displacement distances whenever reliable were also assessed. All the information was reflected in diagrams and pictures.

3 Study of Detritus Flow and Distribution of Materials

Two segments of the coastal fringe were studied: one with vegetation between the land and the sea (natural barrier) and the other with anthropogenic accumulations (man-made barriers), (Fig. 2a, b).

The former has a flat surface with vegetation fringes parallel to the coastline (normal sea level). Although man action is evident, no significant relief disturbances are found. Primary plant cover, however, has been disturbed.

For this reason, the main protection barrier in this segment is a fringe of sea grapes (15–20 m wide), made up by short trees (less than 2–4 m high) with relatively thin trunks (diameter 15–30 cm) and superficial but expanded roots. The rest of the vegetation toward the supratidal–intertidal area is short bushes and grass patches.

Observations in this sector were performed in correspondence with the identified effects along the material (sediments and blocks) flow line, following a sea–land direction. Current direction was from the NW to the SE (azimuth 130°) in correspondence with the prevailing direction of the heavy sustained winds.

In the latter, the artificial barrier is made up by blocks, gravel, sand, and soil arranged with heavy equipment in 2 m high and 5–10 m wide walls that stretch parallel to the coastline, 40–50 m away from it.

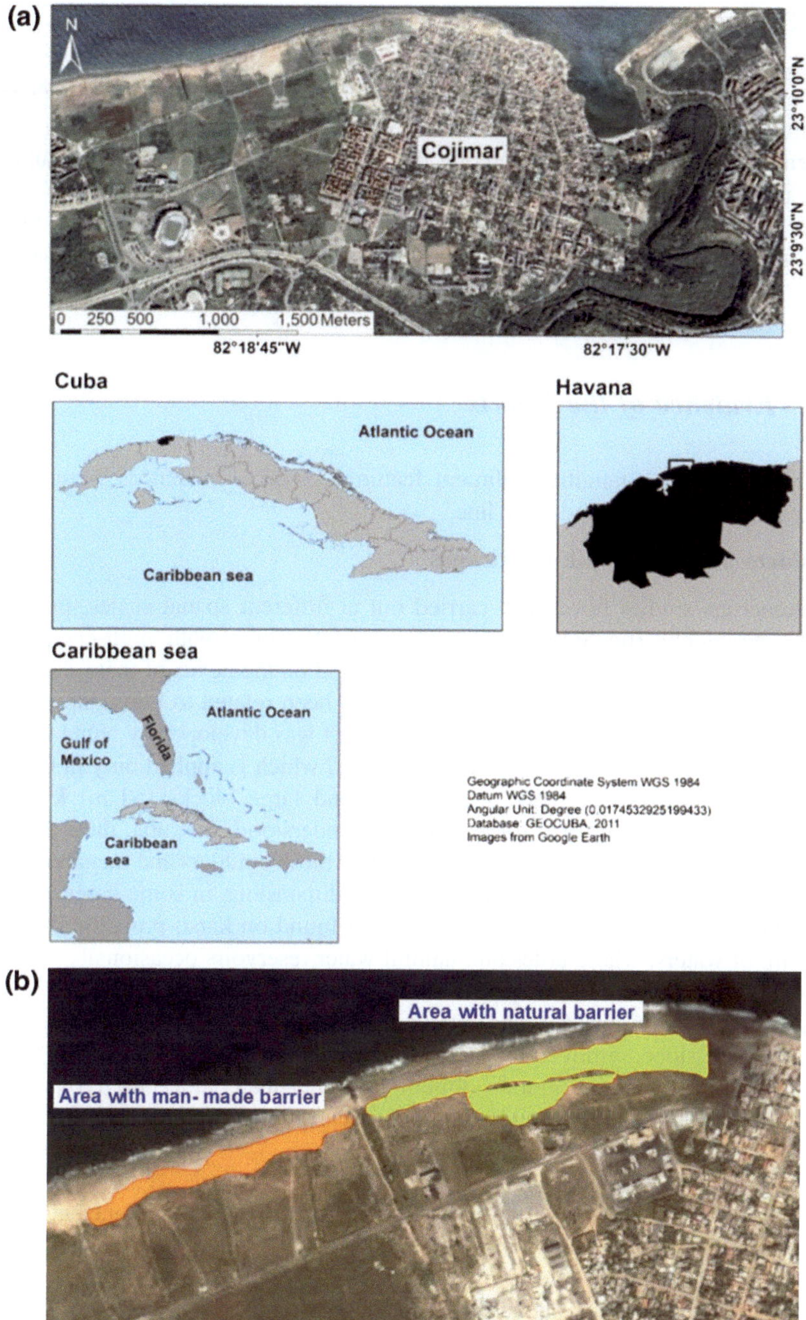

Fig. 2 **a** Location of the town of Cojimar, La Habana, Cuba. **b** Study area, west of the town of Cojimar, in eastern Havana (Google). Area with: man-made barrier and natural barrier

4 Discussion and Results

The effects of this event were verified in different environmental elements such as detritus, coastal rocks, coastal soil, and vegetation, as well as in existing anthropogenic materials. Interactions among these elements show significant regularities, very useful from the methodological and practical points of view.

Below, the effects and possible origin are described. It is evident that the assessment of similar events is required; however, the information presented is quite useful.

4.1 Evidence of the Effects

Based on the most tangible sediment features, physical evidences were assessed following the marine–land flow line.

Boulders associated with hurricanes

Numerous studies have been carried out at different spatial scales, in order to relate the blocks (boulders) deposited in coastal areas with high-energy wave events. Most of these investigations have focused on the nexus of these depositions with tsunamis [7–12], although they have also been related to hurricanes [13–15]. The first work carried out in Cuba on this subject was developed by Núñez-Jiménez [16]; these authors use the term "huracanolitos," which is applied only in Cuba. By boulders, in the present work, megablocks and large blocks (>1 m) left in the coastal zone as a result of tropical storms is understood. In Cuba, there is greater uncertainty about the cause of the movement of these boulders and megaboulders in the coastal littoral [17], due to the significant dimensions in some cases.

In the intertidal zone, dissolution areas are found on karst, probably due to the mixture of waters, where holes and natural water reservoirs occasionally inundated have appeared (referred to as ponds from now on in this work).

During the storm, these ponds receive more boulders from the seabed in shallow areas and the slope, removed by waves (Picture 2). Finally, only these big boulders remain in the ponds, fine grain sediment and other material are driven landwards, and little is trapped in the ponds probably because of the event weakens.

"Stone mill"

Such ponds function like the mills of rounded stones used in ancient times to grind cereals, as they transform clasts and detritus. Once in the ponds, these materials are constantly stirred by water and their particles collide and break. As particles become smaller, are suspended by water, and transported out of the ponds until new events displace them to the supratidal area.

Picture 2 Ponds (left) with subrounded boulders (right) undergoing an intense abrasion process resulting from strong wave action, cold fronts, and tropical storms

Displacement of boulders

Tracking displacement of boulders in this scenario is easy for the trail on the karst (Fig. 3). The effect of karst on the surface of the boulders is evident. The inferior surface is usually clearer and cleaner than the top, as it was exposed. This evidence makes possible to determine displacement, rotation or turning of the boulders in the study area as well as the distance they have been displaced.

Several generations of boulders can be observed: (1) *Ancient boulders.* Dark surfaces, with evidence of weathering, oxides, dark external layers, moss, traces of vegetation, etc.; (2) *Recent boulders.* Clean surfaces, clear colors, no weathering effects, dissolution traces, etc.

These differences make possible to determine the occurrence of past events in the area. It also helps in determining the sequence of boulder accumulation when imbrication patterns are present. Usually, boulder imbrication is evident with

Fig. 3 Big boulder displaced 14 m from site 1 to site 2 (57 m away from the shoreline)

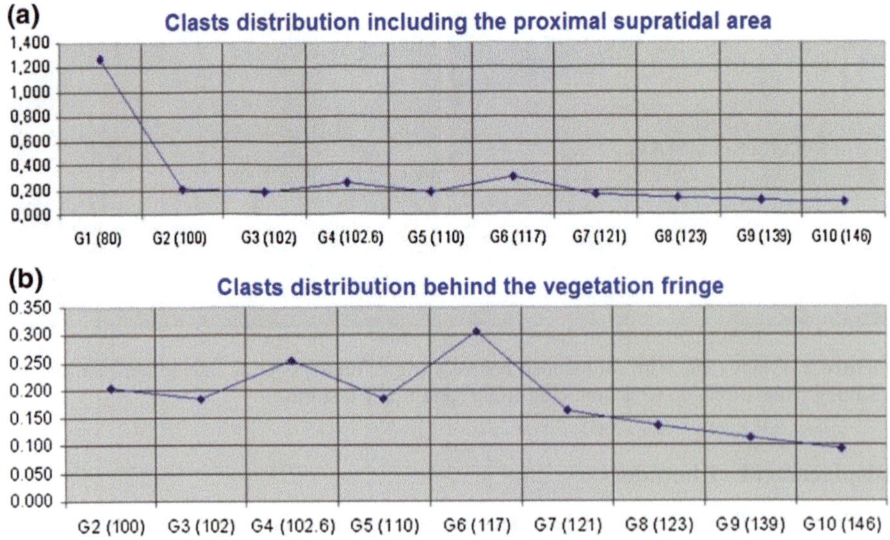

Fig. 4 Boulders prevail in the first 80 m of the intertidal zone (**a**), while their size dramatically decreases behind the vegetation fringe (over 100 m), size increases sometimes in wash over fans (**b**)

45°–60° imbrication angles, with a seaward inclination of the flat surface, in an opposite direction to the flow influence.

It was verified that some blocks were deposited at different distances on the supratidal area, depending on their shape and weight. Their displacing effect on the karst was also expressed in collisions and fragmenting of the edges and acute parts of the rocks. This resulting material could or could not accumulate in the area.

Blocks of over a meter (Fig. 4, see graph) deposited in the supratidal area (karst area 30–40 m wide) by previous events and anthropogenic materials were displaced inland dozens of meters. This phenomenon seems to be cyclical; blocks are displaced several meters every time a storm hits the area, depending on its energy.

Knowledge on the issue will make possible to design a monitoring system for the behavior of this blocks in the coastal area, using markers and GPS, in such a way that processes associated with these events could be assessed.

Soil erosion in the proximal supratidal zone

Soil accumulated over time (soil layer 10–30 cm thick) was removed by the strong flow in the leeward side, widening the intertidal area from some centimeters to over a meter. Some blocks were removed directly from the karst and displaced tens of meters (Picture 3).

The soil removed was partly deposited on a sea grape fringe, and the rest of more clayish material was transported inland. This brought about accumulation of shells, coral fragments, and other biogenic remains.

Picture 3 Block removed from the proximal supratidal zone and displaced tens of meters inland

Vegetation damages

Marine currents and waves breached the plant fringe to allow water flow, trees were uprooted, and branches were broken or twisted, because of the impact of boulders and other solid bodies.

Removal and accumulation of sediments in the vegetation fringe

An important amount of detritus from the sea was trapped in the tangle of sea grape, large blocks (over a meter), boulders, gravel, and sand (generally mixed) accumulated. Some of them were in the soil removed from the edge of the supratidal zone (Picture 4 and Illustration 1). This deposit has a grainy texture, with scarce fine sediments and practically no clay, as it was washed and transported inland.

Picture 4 Blocks trapped in the sea grape fringe. Notice the high erosive effect of water on the karst

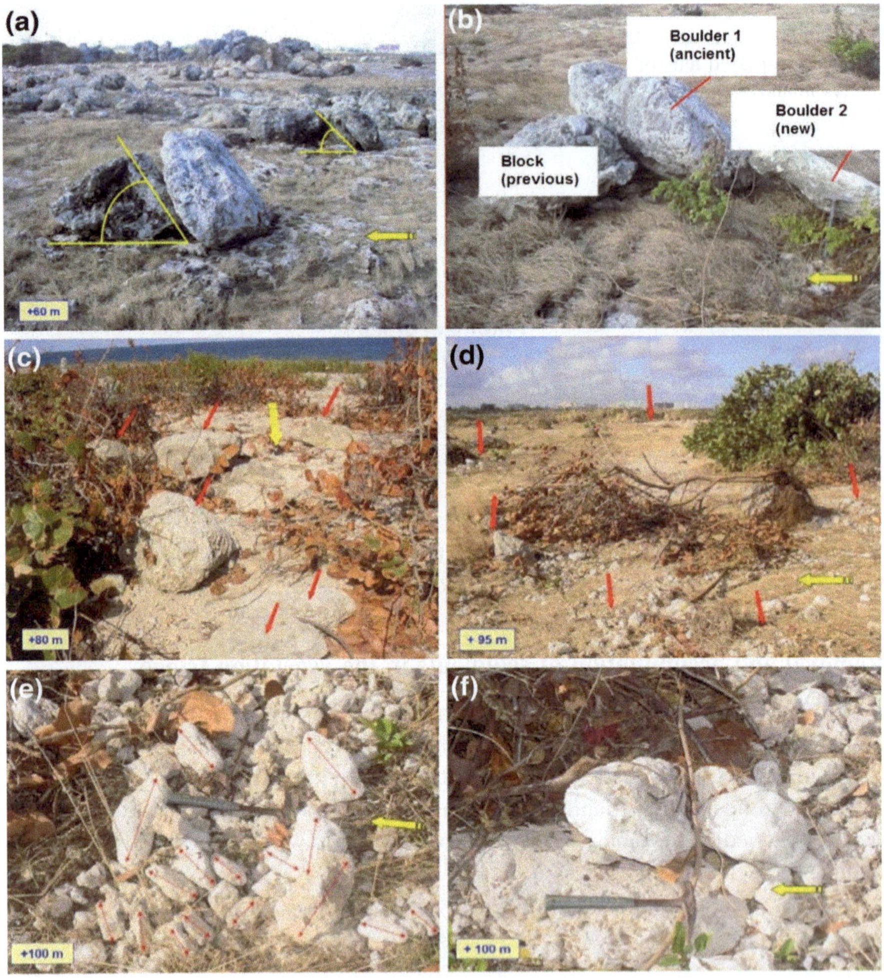

Illustration 1 Sequence of structures and sediment textures resulting from marine flow on the coast (60–100 m from the shoreline). Blocks imbrication and sequence of accumulation of specimens of different generations (above); blocks in the sea grape fringe (middle); orientation and location of blocks regarding size and shape (below)

In the channels formed by the water flow, soil erosion was severe and detritus accumulated only in the banks of the channels, forming ridges (gravel and coarse sand) in the tree trunks and large blocks.

Detrital fan in the natural barrier

The material deposited when currents weakened at end of the channels formed detrital fans of different dimensions. This event favoured the displacement of blocks tens of meters landwards. These deposits are arranged according to their size, but a

Picture 5 Gravel and block fan in the edge of the vegetation fringe toward the protected posterior side

varied mixture of block fragments and gravel is observed. The longest axis of the blocks is transversally oriented regarding current direction, showing that displacement took place by rolling on the ground. There are lines of accumulated material behind and in front of some blocks and relatively large blocks already fixed (Picture 5 and Illustration 1).

Detrital fans in the artificial barrier

Significant detrital fans of thick material (gravel and blocks) were formed in the artificial barrier, with a grainy texture and a peculiar distribution of blocks. These huge clasts rolled over the fan and stopped at the outer edge forming a semicircle. Their wide displacement range inland is of scientific interest. The lower surface and the scarce grass of the area played a role (Picture 6 and Illustration 2).

Picture 6 Detrital fans behind the artificial barrier. Notice radial distribution of blocks landwards

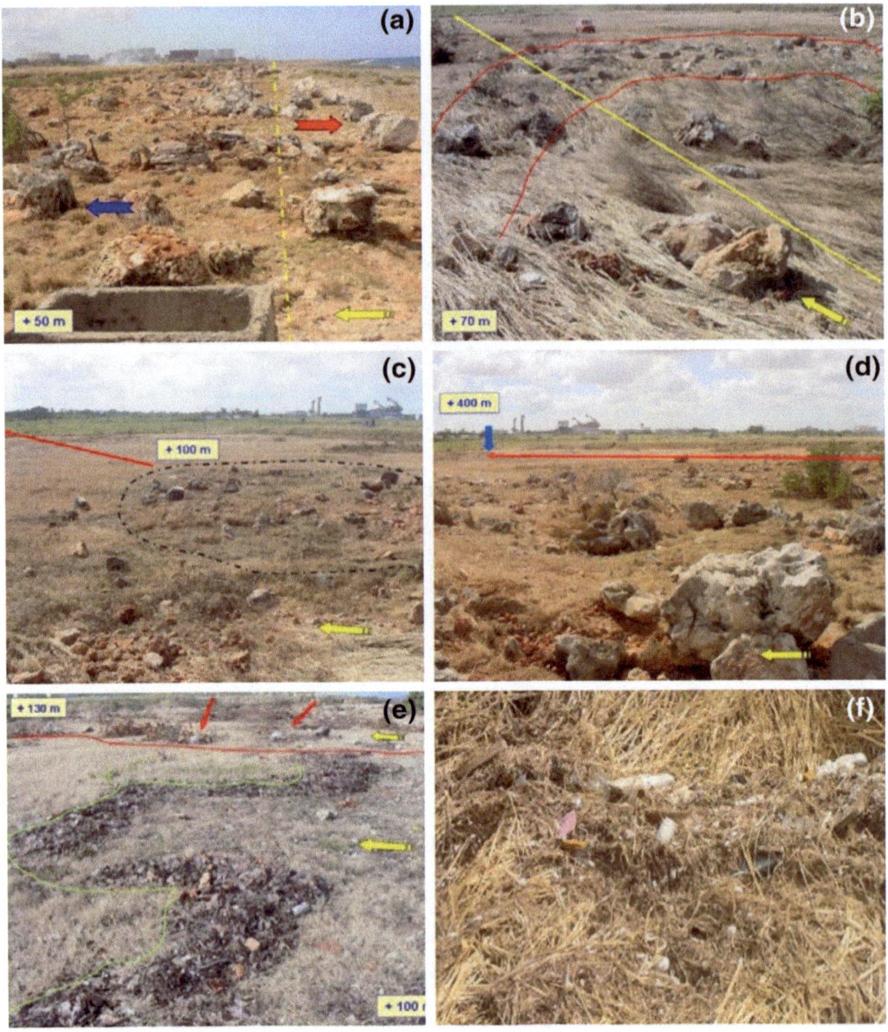

Illustration 2 Wave effect and coarse detrital deposits related to the artificial barrier. Disorganization of blocks resulting from barrier destruction (above and middle). Accumulation of dead leaves filtered by the vegetation fringe in the natural barrier (left below) and garbage accumulation in the distal part during maximum flooding (right below)

Distal deposits

Distal deposits consist of fine sand and clay accumulated in large areas distant from the shore (Illustration 2f). In this case, such deposits had a large amount of light elements which floated in the water flow hundreds of meters inland (200–400 m). In general, sea fans, gorgonians, sponges, and algae, etc., were abundant. Anthropogenic waste material was also significant, mainly city garbage transported by rivers to the coastal area of the city.

Marine flow

The marine flow resulting from the storm was unidirectional (sea to land direction) and determined the direction of the detrital material displaced. It also determined block imbrication, inland displacement, the formation of sediment structure, the zonal distribution of materials, and their selection and accumulation. Unlike tsunami deposits, the flow induced by the storm associated with Hurricane Wilma kept the same direction for almost 72 h, so water of the flooded area did not go back to the sea violently. Otherwise, materials would have shown an opposite direction.

Flow behavior once the storm ended seems to have been influenced by the geology of the coastal area, characterized by a well-developed underground and surface karst. This fact favoured the rapid infiltration of flow water, preventing a higher level of inundation and a violent comeback to the sea, which would have changed sediment distribution patterns.

5 Application of the Study

The evolution of the effects of this event indicates that intensity and behavior of sediment processes are different among coast sectors, depending on the natural or artificial (constructions) protection of the area (Fig. 5).

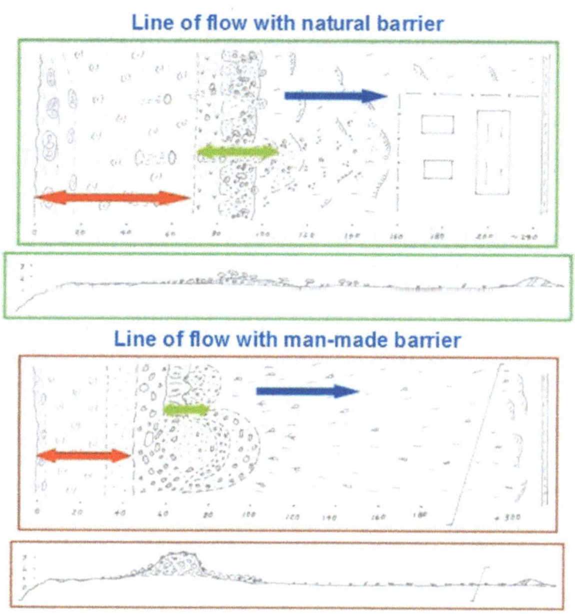

Fig. 5 Notice that material washed from the artificial barrier was accumulated again at a natural optimal distance ranging from 80 to 100 m, similar to the position of detrital accumulation in the natural barrier

5.1 Functional Differences of Barriers

Natural barrier

- The plant fringe protects the supratidal area from waves and wind.
- Blocks or filters marine detritus.
- Its natural recovery does not require resources or infrastructure.

Artificial barrier

- May increase impact with construction materials.
- Increases the discharge of anthropogenic detritus inland.
- Additional resources are required for its reconstruction.

6 Conclusions

1. The features, physical effects, and documented and deduced processes in this work have an important methodological value to assess impact and risks of similar events.
2. The information gathered allows the design of a monitoring program for particular coastal sectors (GPS) to characterize anomalous events.
3. Studying current hurricane effect on sediments could make possible the identification of this type of deposit in the geological records, shedding light on past hurricane effects. Historical records on the issue are extremely limited.
4. The fringe of coastal vegetation acts like a natural filter for storm deposits. The elimination of such fringe would increase negative effects of storms.
5. Constructions on the supratidal area (between the shoreline and the sea grape fringe) are exposed to the impact of hurricanes.

7 Recommendations

- The assessment of the environmental response to an anomalous event, such as Hurricane Wilma, shows the need to take into account the physical impacts to avoid vulnerability of territorial planning projects in coastal areas.
- Increase hurricane impact assessment, including recent geological studies, knowing that historical records on the issue are limited.
- The effects on the natural physical environment should be taken into account when assessing hurricane impact.
- Ban constructions on the proximal supratidal area and influence to remove the existing ones.

References

1. Ballester M, Rubiera J (2007) Temporada ciclónica de 2005 en el Atlántico Norte. http://www.insmet.cu. Last accessed 2007/04/09
2. Klotzbach PJ (2006) Trends in global tropical cyclone activity over the past twenty years (1986–2005). Geophys Res Lett 33:L10805
3. Peros MC, Gregory B, Matos F, Reinhardt E, Desloges J (2015) Late-holocene record of lagoon evolution, climate change, and hurricane activity from southeastern Cuba. Holocene 25(9):1483–1497
4. Arche A (1992) Sedimentología. Volumen 1. Consejo Superior de Investigaciones Científicas, Madrid, España
5. Vera-Torre JA (1994) Estratigrafía, principios y métodos. Editorial Ruedas, S. L. Madrid, España
6. Dabrio CJ (1992) Playas e islas barrera-lagoon, en: Arche A.: Sedimentología. Volumen 1. Consejo Superior de Investigaciones Científicas, Madrid, España
7. Scheffers A (2002) Paleotsunami in the Caribbean: field evidences and datings from Aruba, Curaçao and Bonaire, Essener Geographische Arbeiten, 33
8. Scheffers A, Kelletat D (2003) Sedimentologic and geomorphologic tsunami imprints worldwide—a review. Earth Sci Rev 63:83–92
9. Goto K, Chavanich SA, Imamura F, Kunthasap P, Matsui T, Minoura K, Sugawara D, Yanagisawa H (2007) Distribution, origin and transport process of boulders transported by the 2004 Indian Ocean tsunami at Pakarang Cape, Thailand. Sediment Geol 202:821
10. Goto K, Kawana T, Imamura F (2010) Historical and geological evidences of boulders deposited by tsunamis, southern Ryukyu Islands, Japan. Earth Sci Rev 102:77–99
11. Engel M, Brückner H, Wennrich V, Scheffers A, Kelletat D, Vött A, Schäbitz F, Daut G, Willershäuser T, May SM (2010) The back-barrier record of eastern Bonaire (Netherlands Antilles): new insights into the palaeo-tsunami history of the southern Caribbean. Sediment Geol 231:14–30
12. Engel M, May SM (2012) Bonaire's boulder fields revisited: evidence for holocene tsunami impact on the Leeward Antilles. Quatern Sci Rev 54:126–141
13. Goto K, Miyagi K, Kawana T, Takahashi J, Imamura F (2011) Emplacement and movement of boulders by known storm waves—field evidence from the Okinawa Islands, Japan. Mar Geol 283:66–78
14. May SM, Brill D, Engel M, Scheffers A, Pint A, Opitz S, Wennrich V, Squire P, Kelletat D, Brückner H (2015) Traces of historical tropical cyclones and tsunamis in the Ashburton Delta (NW Australia). Sedimentology 62:1546–1572
15. Engel M, May SM, Brill D, Cuadra C, Mahar A, Lagmay F, Santiago J, Kenneth J, Hoffmeister D, Brückner H (2016) Boulders shifted during Supertyphoon Haiyan (7–9 Nov 2013)–observations from eastern Samar (Philippines). Geophys Res Abs 18. (EGU2016-15390)
16. Núñez-Jiménez A, Viña N, Acevedo M, Mateo J, Iturralde-Vinent M (1988) Cuevas y carsos. Ed. Científico-Técnica, La Habana
17. Matos F (2017) Boulders in Cuba: hurricanes or tsunamis? Project Paleotempestología en Cuba: variabilidad espacio–temporal de la actividad de huracanes usando proxy data. Data, Mar 2017. https://doi.org/10.13140/rg.2.2.17717.88809, www.researchgate.net

Mathematical Modeling of Phosphorus Dynamics in Aquatic Ecosystems

Maibelin Castillo-Alvarez, Rolando Cárdenas, Roberto González-de Zayas, Yanelis Estrada-Hernández, Julio Antonio Lestayo, Dailé Ávila-Alonso and Lorgio Batar

Abstract A basic framework for the elaboration of a model of the dynamics of phosphorus in Cuban coastal ecosystems is presented. We start from a zero-dimensional model, which includes several physical, chemical, and biological processes depending only on time. In this, first a relatively isolated ecosystem was considered, and then polluting loads from runoff were included. Then, the vertical space variable was introduced, given the importance of the interaction of phosphorus with the bottom in many coastal ecosystems. This resulted in a one-dimensional model of predominantly thermo-hydrodynamic model, coupled with a model of radiative transfer in atmosphere and ocean. The thermo-hydrodynamic frame of the model consists of a system of three partial differential equations whose solutions were found and turned out to be well behaved from a mathematical point of view. A preliminary application to a case study was done and analyzed. They were able to lay the foundations for the preparation of a Cuban hydrodynamic model.

Keywords Phosphorus · Aquatic ecosystems · Differential equation Biogeochemical processes

1 Introduction

All forms of life known on Earth contain at least six chemical elements: carbon (C), hydrogen (H), oxygen (O), nitrogen (N), phosphorus (P), and sulfur (S). However, those that limit the growth of organisms are usually N or P. In the case of aquatic

M. Castillo-Alvarez (✉) · R. G. Zayas · J. A. Lestayo
Coastal Ecosystem Research Center (CIEC), Los Almácigos Avenue, Cayo Coco, Morón, Ciego de Ávila, Cuba
e-mail: maibelinca7@gmail.com

R. Cárdenas · Y. Estrada-Hernández · D. Ávila-Alonso · L. Batar
Central University "Marta Abreu" of Las Villas, Road Camajuaní km. 5.5, Santa Clara, Villa Clara, Cuba

R. Cárdenas et al. (eds.), *Proceedings of the 2nd International Conference on BioGeoSciences*, https://doi.org/10.1007/978-3-030-04233-2_11

ecosystems, nitrogen is often the limiting factor in salt water and phosphorus in freshwater [1, 2]. However, this depends on a multitude of physical, chemical, and biological processes to which both elements (and their compounds) are subject.

Another aspect to consider is the diversity of compounds that phosphorus forms in aquatic environments. They are usually classified into four different categories corresponding to their physical–chemical properties, such as particulate refractory organic phosphorus, particulate labile organic phosphorus, dissolved organic phosphorus and total phosphate (PO_{4t}) or soluble reactive phosphorus (PRS). These are not totally independent of each other; currently, processes are known that transform the first three categories into PRS. From the biological point of view, it is very important if we take into account that the PRS is the way in which phosphorus is mostly assimilated by algae [3].

The increase in phosphorus concentrations in aquatic ecosystems has led to episodes of eutrophication, for example, in agricultural regions with extensive use of fertilizers. The decrease of this in coastal zones, by damming rivers or drought, has led to notable alterations in the trophic assembly [4, 5].

Models of marine ecosystems, which include the interaction of physical and biological processes, are the only tool to investigate on a large-scale aspect related to the process of important current problems, such as the understanding and quantification of biogeochemical cycles in the ocean. This is the reason for the recent trend toward the development of three-dimensional physical–biological models for application at global or at least basin scales. It is reasonable to support this work through research using one-dimensional models that offer the opportunity to study the suitability of different structures of ecosystem models, to adapt mathematical descriptions of biological or chemical processes and the consequences of special parameter options with a modest computational effort compared to different three-dimensional models [6].

Higher resolution and dimensionality are increasingly common as computer resources improve, but three-dimensional modeling should not always be the default. Sometimes, three dimensions are necessary to simulate important physical dynamics, but implementing a three-dimensional model of reasonable resolution is costly from a computational point of view, which limits the options available for parameter estimation, sensitivity analysis, and uncertainty analysis [6].

At the international level, various investigations have been carried out on the biogeochemical behavior of phosphorus in different aquatic ecosystems. The objective of these studies is to have a broader knowledge of the dynamics of this nutrient. Given its complexity, it is considered that the deductions obtained do not reflect very realistic results.

In Cuba, there are few studies related to the behavior of phosphorus in aquatic ecosystems. The numerical models developed by the Land–Ocean Interactions in the Coastal Zone (LOICZ) has been applied in Larga Lagoon [7]. These models describe the dynamics of the biogeochemical processes in the coastal zone, both at local, regional, and global scales and are related to the cycles of C, N, and P [8]. LOICZ was designed to describe the hydrological equilibrium ocean–coastal zone from the balances of water, salt, and nutrients [9]. It is very important to develop mathematically treatable models, which can be implemented in the country. It is very important to develop mathematically treatable models, which can be implemented in the country, which implies a reduction in costs.

The modeling of the phosphorus dynamics in aquatic ecosystems has a high degree of complexity due to the multitude of physical, chemical, biological, and geological processes to which this chemical element is subjected. The foregoing illustrates the importance of obtaining predictions of phosphorus dynamics in coastal ecosystems.

That is why the objective of this paper is to propose a model of the phosphorus dynamics for Cuban coastal ecosystems.

2 Materials and Methods

The following conceptual model (Fig. 1) represents a series of processes and interactions between the biotic and abiotic environment typical of marine ecosystems.

Phosphorus in seawater is found in living organisms or as dissolved inorganic phosphorus (DIP), dissolved organic phosphorus (DOP), and particulate organic phosphorus (POP). In seawater, where cell concentrations are lower than in lakes, the dissolved organic fraction (which comes from cell exudates) may be smaller than the inorganic fraction. Uptake by primary producers and bacteria is responsible for the low phosphate concentration typical of surface waters. The release of phosphate directly from microalgae is very low, and the amounts produced by hydrolysis of DOP are very small. Some phosphate is excreted by bacteria, and DIP is also provided by microbial hydrolysis of the esters of DOP. This lysis is a very rapid process that limits the persistence of DOP to a few hours. By far, phosphate is

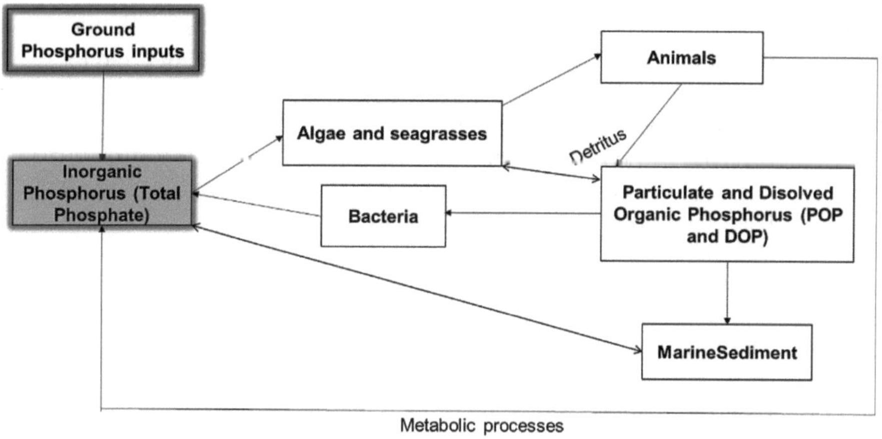

Fig. 1 Phosphorus dynamics in aquatic ecosystems

principally regenerated by the decay of particulate organic phosphorus (POP) and by animals. Some of the POP is released as DIP as particles decay in the water column, but some settle onto sediments. In the sediments, further degradation of settled organic P to DIP can take place and some of this DIP is precipitated or adsorbed [10].

2.1 Zero-Dimensional Model

In order to address the dynamics of the PRS, we start from the kinetic equation of the PRS (5.4.7) that appears in [3]:

$$\begin{aligned}\frac{\partial \mathrm{PO_{4t}}}{\partial t} = & \sum_{x=c,d,g} (\mathrm{FPI}_x \cdot \mathrm{BM}_x + \mathrm{FDIP} \cdot \mathrm{PR}_x - \mathrm{P}_x)\mathrm{APC} \cdot \mathrm{B}_x \\ & + \mathrm{K_{DOP}} \cdot \mathrm{DOP} + \frac{\partial(\mathrm{WS_{TSS}} \cdot \mathrm{PO_{4p}})}{\partial z} + \frac{\mathrm{BFPO_{4d}}}{\Delta z} + \frac{\mathrm{WPO_{4t}}}{V},\end{aligned} \tag{1}$$

where

$\mathrm{PO_{4t}}$	Total phosphate (gP m^{-3}), that is, the sum of dissolved and absorbed phosphate,
t	Time (day),
$\mathrm{PO_{4p}}$	Absorbed phosphate (gP m^{-3}),
FPI_x	Fraction of phosphorus metabolized by algae of group x, carried to inorganic phosphorus,
c	Group of algae "cyanobacteria,"
d	Group of algae "diatoms,"
g	Group of "green–blue" algae,
BM_x	Basal metabolism of group x of algae (gP m^{-3} day),
FDIP	Fraction of phosphorus predated carried to inorganic phosphorus,
PR_x	Predation rate of phosphorus by group x of algae (gP m^{-3} day),
P_x	Growth rate of group x of algae (day^{-1}),
APC	Average phosphorus/carbon ratio in all algae groups,
B_x	Biomass of group x of algae (gC m^{-3}),
K_{DOP}	Hydrolysis rate of dissolved organic phosphorus (day^{-1}),
DOP	Dissolved organic phosphorus (gP m^{-3}),
$\mathrm{WT_{TSS}}$	Sedimentation velocity (m day^{-1}),
z	Vertical coordinate (m),
$\mathrm{BFPO_{4d}}$	Exchange of phosphate flow between sediment and water (gP m^{-2}day),
Δz	The thickness of the bottom layer in which phosphate exchange occurs (m),
$\mathrm{WPO_{4t}}$	External charges (additions) of total phosphate (gP day^{-1}),
V	Volume (m^3).

2.1.1 Considerations

First approximation

A relatively isolated ecosystem is assumed, which justifies neglecting the last term of Eq. (1). In order to increase the mathematical treatability, it is proposed to consider as a constant K the other terms of the right side, except for the term of the derivative.

Second approximation

It is considered to include in the model the external addition of phosphate by runoff from watersheds. Equation (1) can then be written:

$$\frac{\partial(\mathrm{PRS})}{\partial t} = K_1 + K_2 \frac{\partial}{\partial z}(\mathrm{PRS}). \tag{2}$$

2.2 *One-Dimensional Model*

Given the importance of the interaction of phosphorus with the bottom of the different bodies of water, it was considered to include the vertical spatial coordinate, resulting in a one-dimensional model. To accomplish this, a pelagic biophysical model for nitrogen was adapted to a coastal model for phosphorus. The starting model is a two-level second-order turbulence closure model developed by Mellor and Yamada [11, 12], which is described by a system of partial differential equations for the moment and the temperature:

$$\frac{\partial u(z,t)}{\partial t} = \frac{\partial\left[(A_{\mathrm{M}} + \upsilon_{\mathrm{M}})\frac{\partial u(z,t)}{\partial z}\right]}{\partial z} + fv(z,t), \tag{3}$$

$$\frac{\partial v(z,t)}{\partial t} = \frac{\partial\left[(A_{\mathrm{M}} + \upsilon_{\mathrm{M}})\frac{\partial v(z,t)}{\partial z}\right]}{\partial z} - fu(z,t), \tag{4}$$

$$\frac{\partial T(z,t)}{\partial t} = \frac{\partial\left[(A_{\mathrm{H}} + \upsilon_{\mathrm{H}})\frac{\partial T(z,t)}{\partial z}\right]}{\partial z} + \frac{1}{\rho c_{\mathrm{p}}}\frac{\partial I(z)}{\partial z}, \tag{5}$$

where

u and v	The horizontal components of the speed in the east and north directions, respectively (m s^{-1}),
T	Temperature (°C),
t	Time (s),

z	The vertical coordinate (m),
f	Coriolis parameter,
ρ	Density of sea water (Kg m^{-3}),
c_p	The specific heat capacity of seawater under normal conditions (J $Kg^{-1}(^{o}C)^{-1}$),
I	The intensity of solar radiation at depth z (W m^{-2}),
A_M and A_H	Vertical swirl diffusivities are dependent on depth, for the moment and heat (m^2 s^{-1}),
υ_M and υ_H	The viscosity and molecular diffusivity, respectively (m^2 s^{-1}).

The proposed model consists of three differential equations in partial derivatives of the second order:

$$\frac{\partial u(z,t)}{\partial t} = K_3 \frac{\partial^2 u(z,t)}{\partial z^2}, \tag{6}$$

$$\frac{\partial v(z,t)}{\partial t} = K_3 \frac{\partial^2 v(z,t)}{\partial z^2}, \tag{7}$$

$$\frac{\partial T(z,t)}{\partial t} = K_4 \frac{\partial^2 T(z,t)}{\partial z^2} + \frac{1}{\rho c_p} \frac{\partial I(z)}{\partial z}. \tag{8}$$

2.2.1 Considerations

First approximation

The force of Coriolis was rule out, because the acceleration that this implies is very small, and therefore, its influence on the velocity of the fluid elements is not very noticeable for short vertical trajectories, as is the case of the coastal areas.

Second approximation

It was considered that the swirl diffusivities for the moment and the temperature (A_M and A_H) and the viscosity and molecular diffusivity (υ_M and υ_H) are constant; therefore,

$$A_M + \upsilon_M = K_3, \tag{9}$$

$$A_H + \upsilon_H = K_4. \tag{10}$$

2.2.2 Case Study: Ana Maria Gulf

The Ana Maria Gulf (Fig. 2) [13], located to the south of the provinces of Sancti Spíritus, Ciego de Ávila, and Camagüey, has great economic importance for our country, especially for being a transcendental area in terms of fishing and tourism. The following sections show aspects considered to apply the one-dimensional model proposal. In most cases, continuity was pursued with previous work with the Ana Maria gulf carried out in the author's group [14].

Depth

The average depth was obtained from the sampling stations considered in Alvarez-Salgueiro [14], which resulted in 11.4 m.

Time

A cycle of one full year was considered, selecting to evaluate the solutions of the speeds those times that mark the changes of the seasons: the solstices and equinoxes (Table 1).

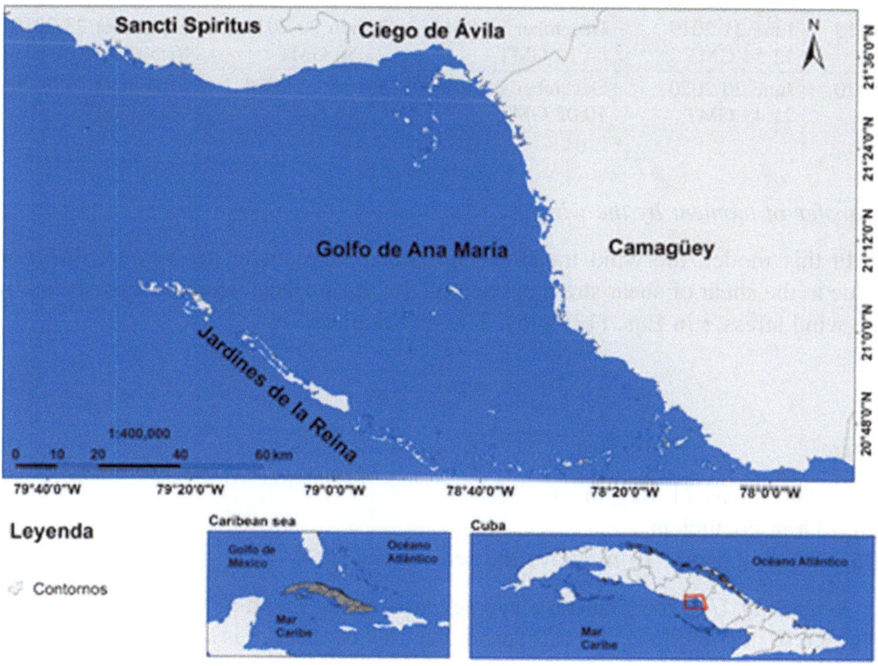

Fig. 2 Ana Maria gulf, Cuba

Table 1 Dates and times of the solstices and equinoxes 2010–2020 [14]

Year	Summer solstice	Winter solstice	Spring equinox	Autumn equinox
2010	June 21 2010 11:28 GMT	December 21 2010 23:38 GMT	March 20 2010 17:32 GMT	September 23 2010 03:09 GMT
2011	June 21 2011 17:16 GMT	December 22 2011 05:30 GMT	March 20 2011 23:21 GMT	September 23 2011 09:04 GMT
2012	June 20 2012 23:09 GMT	December 21 2012 11:11 GMT	March 20 2012 05:14 GMT	September 22 2012 14:49 GMT
2013	June 21 2013 05:04 GMT	December 21 2013 17:11 GMT	March 20 2013 11:02 GMT	September 22 2013 20:44 GMT
2014	June 21 2014 10:51 GMT	December 21 2014 23:03 GMT	March 20 2014 16:57 GMT	September 23 2014 02:29 GMT
2015	June 21 2015 16:38 GMT	December 22 2015 04:38 GMT	March 20 2015 22:45 GMT	September 23 2015 08:20 GMT
2016	June 20 2016 22:34 GMT	December 21 2016 10:44 GMT	March 20 2016 04:30 GMT	September 22 2016 14:21 GMT
2017	June 21 2017 04:24 GMT	December 21 2017 16:28 GMT	March 20 2017 10:28 GMT	September 22 2017 20:02 GMT
2018	June 21 2018 10:07 GMT	December 21 2018 22:22 GMT	March 20 2018 16:15 GMT	September 23 2018 0154 GMT
2019	June 21 2019 15:54 GMT	December 22 2019 04:19 GMT	March 20 2019 21:58 GMT	September 23 2019 07:50 GMT
2020	June 20 2020 21:43 GMT	December 21 2020 10:02 GMT	March 20 2020 03:49 GMT	September 22 2020 13:30 GMT

Transfer of moment by the wind

In this model, the wind transmits horizontal moment in the vertical direction z due to the shear or shear stress exerted on the sea surface, which is represented by the wind stress τ in Eqs. (13)–(15). This is calculated by:

$$\tau = \rho_a C_D U_{10}^2, \tag{11}$$

where

ρ_a Air density (1.3 kg m^{-3}),
C_D Drag coefficient,
U_{10} Wind speed at 10 m above the sea surface (m s^{-1}).

Regarding U_{10}, the planetary average was used: 7.4 m s^{-1} [16]. According to Fig. 4.6 of this reference, for this speed, the drag coefficient is approximately 0.001.

Swirl diffusivity for the moment (A_M) and viscosity (υ_M)

These parameters are very important, since they influence the transfer of moment in the vertical direction and therefore in the regime of marine current that is established. The values taken are:

υ_M 1.34×10^{-5} m^2 s^{-1} [17],
A_M 1.3×10^{-4} m^2 s^{-1}, average diffusivity of swirls [16].

3 Results and Discussion

3.1 Zero-Dimensional Model

When solving (2), by the method that appears in [18, p. 248], results:

$$\mathrm{PRS} = K_1 t + \mathrm{PRS}_0 \cdot \mathrm{PRS} = K_1 t + \mathrm{PRS}_0. \tag{12}$$

The solution obtained shows a linear dependence of the PRS with time.

3.1.1 Analysis of the Zero-Dimensional Model

K_1 could be considered approximately constant in very specific situations in which there could be a constant difference between the terms that provide PRS (the positive ones) and those that extract it (the negative ones). This frames the limits of the application of this model and defines a further step to follow: to determine under what conditions this record of K_1 could be fulfilled.

K_1 could reflect reality for brief periods of time (days) when the dynamics of phosphorus could vary slowly in a given aquatic ecosystem. However, it is expected that this will not be accomplished for large periods of time (rain and drought seasons).

3.2 One-Dimensional Model

To solve the system of differential equations proposed, leaving for further studies the development of the temperature equation due to its high degree of complexity, by the method of separation of variables [19], the following initial and boundary conditions are proposed:

For $u(z, t)$:

$$\left.\begin{aligned} u(z,0) &= \frac{\tau}{\rho K_3} f(l+z) + U_0, \\ \left(\frac{\partial u(z,t)}{\partial z}\right)_{z=0} &> 0, \\ u(0,t) &= \frac{\tau l}{\rho K_3}, \\ u(l,t) &= 0. \end{aligned}\right\} \tag{13}$$

For $v(z, t)$:

$$\left.\begin{aligned} v(z,0) &= \frac{\tau}{\rho K_3} f(l+z) + V_0, \\ \left(\frac{\partial v(z,t)}{\partial z}\right)_{z=0} &> 0, \\ v(0,t) &= \frac{\tau l}{\rho K_3}, \\ v(l,t) &= 0. \end{aligned}\right\} \tag{14}$$

where

u_0, v_0 Initial speeds (m s^{-1}),
T_0 Room temperature (°C),
τ Component of the wind stress on the surface of the water (Kg m^{-1} s^{-2}),
f Calibration factor.

The solutions of the system are:

$$u(z,t) = v(z,t) = \sum_{n=1}^{\infty} \left[-\frac{4\tau l(-1+(-1)^n)}{\rho K_3 n^2 \pi^2} e^{-\left(\frac{n\pi}{l}\right)^2 K_3 t} \operatorname{sen} \frac{n\pi}{l} z \right] + \frac{f\tau(l+z)}{\rho K_3}. \tag{15}$$

3.2.1 Analysis of the Mathematical Properties of the One-Dimensional Model

The terms with order number n even in the series of solutions are canceled; then, the solutions are:

$$u(z,t) = v(z,t) = \sum_{n=0}^{\infty} \left\{ \frac{A}{(2n+1)^2} e^{-\left(\frac{(2n+1)\pi}{l}\right)^2 K_3 t} \operatorname{sen} \frac{(2n+1)\pi}{l} z \right\} + \frac{f\tau(l+z)}{\rho K_3}. \tag{16}$$

where

$$A = \frac{8\tau l}{\rho K_3 \pi^2} \tag{17}$$

The series of Eq. (16) is continuous and converges uniformly, and this is demonstrated by the Weierstrass criterion, which means that the solution can be directly substituted in the initial equations and can be derived (or integrated) term by term.

3.2.2 Analysis of the Physical Properties of the One-Dimensional Model

The series of Eq. (16) is convergent and hyper-operative, so they tend to a certain value. This is of paramount importance because it allows the series to be truncated, since infinite terms are not physically useful. The series was truncated for $n = 1$, because the sum of the first two terms of the series will be greater than the sum of the other terms, the latter being the absolute error that is committed when performing the truncation. The solutions are:

$$u(z,t) = v(z,t) = Ae^{-\left(\frac{\pi}{l}\right)^2 K_3 t}\text{sen}\frac{\pi}{l}z + \frac{1}{9}Ae^{-\left(\frac{3\pi}{l}\right)^2 K_3 t}\text{sen}\frac{3\pi}{l}z + \frac{f\tau(l+z)}{\rho K_3}, \tag{18}$$

where the absolute error is 4.78×10^{-10} and the relative error is 0.153% (referred to the absolute value of the sum of the first two terms).

3.3 *Marine Current Pattern Obtained*

With the previous data, we proceeded to evaluate the solutions for the speeds. For this, the calibration factor f was assigned a value that reproduced the average velocity in the surface of the Ana Maria gulf, which was taken as 0.15 m s^{-1}, according to data reported in [20]. Figure 3 shows the speeds obtained, for the summer equinox (September 22), since it is the date closest to the sampling campaign whose data were used in [14]. The velocity pattern obtained for the other dates considered is similar.

This velocity pattern is typical of stationary fluid circulation regimes and indicates that time-dependent terms (those in the series) are very small relative to the independent term of time. This pattern has been observed frequently in nature, and more complex patterns that reflect a more complex circulation could be obtained considering the swirl diffusivity and the viscosity as functions and not as constants, which will be considered in later studies.

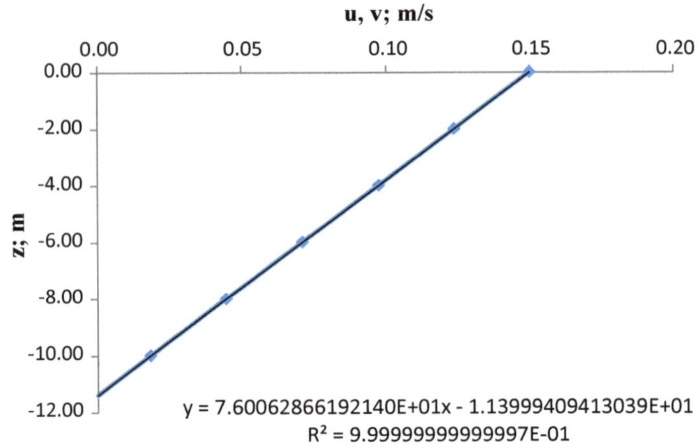

Fig. 3 Speeds of ocean currents at different depths

Figure 3 shows how, at greater depth, the velocities of the currents will decrease to their minimum value at approximately 11.4 m depth. It is a very realistic result, because on the surface the currents get the maximum speed, given the interaction of the wind with it.

4 Conclusions

The conceptual bases for the elaboration of a model of the phosphorus dynamics in the Cuban coastal ecosystems have been presented. For the zero-dimensional case, a linear analytical solution was obtained to the complicated model presented in [3]. This solution can represent situations in the ecosystems in temporal scales of the order of the days, but it is expected that the greater variability that implies the change of seasons (rain drought) responds to more complicated solutions. Subsequently, a one-dimensional model with the inclusion of the vertical spatial variable was presented. The analytical solutions of the corresponding system of equations in partial derivatives were obtained, and a good behavior of these was verified from the mathematical point of view. The analysis of the physical applicability of the solutions was started, through the application to the Gulf of Ana María as a case study, proposing perspectives to make the one-dimensional model more comprehensive.

References

1. Abell JM, Ozkundakci D, Hamilton DP (2010) Nitrogen and phosphorus limitation of phytoplankton growth in New Zealand Lakes: implications for eutrophication control. Ecosystems 13(7):966–977
2. Elser JJ, Marzolf ER, Goldman CR (1990) Phosphorus and nitrogen limitation of phytoplankton growth in the freshwaters of North America: a review and critique of experimental enrichments. Can J Fish Aquat Sci 47:1468–1477
3. Ji Z-G (2008) Hydrodynamics and water quality. Wiley Interscience
4. Márquez A, Senior W, Martínez G, Gonzalez Á (2007) Concentraciones de nitrógeno y fósforo en sedimentos recientes de la laguna Los Patos, Estado Sucre, Venezuela
5. Montalvo JF et al (2010) Compuestos de nitrógeno y fósforo en las aguas superficiales de tres zonas de la plataforma marina cubana. Ser Oceanol 7
6. Robson BJ (2014) State of the art in modelling of phosphorus in aquatic systems: review, criticisms and commentary. Environ Model Softw 61:339–359
7. González-De Zayas R, Merino-Ibarra M, Soto-Jiménez MF, Castillo-Sandoval FS (2013) Biogeochemical responses to nutrient inputs in a Cuban coastal lagoon: runoff, anthropogenic, and groundwater sources. Springer Environ Monit Assess 185(12):10101–10114
8. Montalvo JF, Loza S (2006) Flujos de materiales conservativos y no conservativos en la Bahía de Jigüey (Archipiélago Sabana-Camagüey, Cuba) y el océano. Ser Oceanol 2:1–2
9. López-Monroy F, Troccoli-Ghinaglia L (2017) Modelaje de la interaccón entre la laguna costera tropical Los Mártires (Isla de Margarita, Venezuela) y el Mar Caribe Adyacente. Ciencias Básicas y Tecnología 29:534–545
10. Valiela I (2013) Marine ecological processes. Springer Science & Business Media
11. Mellor GL, Yamada T (1974) A hierarchy of turbulence closure models for planetary boundary layers. J Appl Meteorol 13:1791–1806
12. Mellor GL, Yamada T (1982) Development of a turbulence closure model for geophysical fluid problems. Rev Geophys Space Phys 20:851–875
13. Tageo (ed) (2018) Available (17/04/2018): http://www.tageo.com/index-e-cu-v-00-d-m2276714.htm
14. Alvarez-Salgueiro J (2015) Habitabilidad Primaria del Fitoplancton en el Golfo de Ana María. Departamento de Física, Universidad Central "Marta Abreu" de Las Villas Santa Clara
15. Goddes TW (ed) (2018) Solstice and equinox dates 2010 to 2020. Available (18/01/18): http://www.thewhitegoddess.co.uk/the_wheel_of_the_year/solstice_and_equinox_dates_2010_to_2020.asp (Online)
16. Stewart RH (2006) Introduction to physical oceanography, Sept 2006 edition ed. Texas A & M University
17. Mellor GL, Durbin PA (1975) The structure and dynamics of the ocean surface mixed layer. J Phys Oceanogr 5(7):18–728
18. Elgoltz L (1983) Ecuaciones diferenciales y cálculo variacional, 3ra edn. Editorial Mir, Moscú
19. Tijonov AA, Samarsky A (1972) Ecuaciones de la física matematica. Mir, Moscú
20. Arriaza L et al (2008) Corrientes Marinas estimadas en la plataforma suroriental cubana. Ser Oceanol 4

Mozambican Adsorbents for Zinc (II) Removal in Aqueous Solutions

Julio Omar Prieto-García, Esnaider Rodríguez Suarez, Noor Jehan Gulamussen and Ángel Mollineda Trujillo

Abstract The chapter presents a study of the adsorption capacity of zinc ions from aqueous solutions in natural adsorbents (bagasse ash of red variety sugarcane, coconut shell, activated carbon, and bentonite) from Mozambique. As a heavy metal, zinc has maximum allowable concentration in the drinking water of 5 mg/L according to the Mozambican norm. The environmental pollution caused by the discharge of effluents with contents outside the dumping regulations of Mozambique can be reduced by the use of local materials. According to the results of kinetic studies, the models which better adjust were the pseudo-second order which indicate a physical-type adsorption. Bagasse ash of red variety sugarcane presented the higher adsorption capacity while coal had the lower capacity.

Keywords Adsorbents · Zinc · Adsorption · Kinetic model

1 Introduction

Zn (II) ion is an essential micronutrient and, in general, is considered one of the elements less dangerous. For humans, the feeding sources of zinc are mainly oysters, animal's livers, beer yeasts, meat, and vegetables [5]. The concentration of

J. O. Prieto-García (✉)
Departamento Licenciatura en Química, Facultad de Química y Farmacia, Universidad Central "Marta Abreu" de las Villas, Carretera a Camajuaní km 5 ½, Santa Clara, Villa Clara, Cuba
e-mail: omarpg@uclv.edu.cu

E. R. Suarez · N. J. Gulamussen
Departamento de Química, Faculdade de Ciencias, Universidade Eduardo Mondlane, Avenida Julius Nyerere, Campus Universitario principal, Maputo, Mozambique

Á. M. Trujillo
Centro de Investigaciones Agropecuarias (CIAP), Facultad de Ciencias Agropecuarias, Universidad Central "Marta Abreu" de las Villas, Carretera a Camajuaní km 5 ½, Santa Clara, Villa Clara, Cuba

R. Cárdenas et al. (eds.), *Proceedings of the 2nd International Conference on BioGeoSciences*, https://doi.org/10.1007/978-3-030-04233-2_12

Zinc in drinking water is low and varies from 0.01 to 1 mg/L. If there are favorable conditions, the gastrointestinal absorption is 20–30% [4]. Thus it is favored by cysteine, methionine and histidine amino acids, sugars like fructose and lactose, and vitamin C. On the contrary, it is hindered by high doses of phosphorus, copper, manganese, iron, and tin. Zinc accumulates in the liver, pancreas, kidneys, and prostate [2].

The presence of zinc in a wide variety of enzymes demonstrates its important role in metabolism. The Food and Nutrition Board of the National Academy of Sciences recommends a daily intake of 15 mg of this element for adults. Its lack causes growth decline, marked hypogonadism, and rough and dry skin [2].

Despite being an essential element, at high dosages can be toxic for humans. Among others, toxicity symptoms include vomiting, dehydration, electrolyte imbalance, abdominal pain, lethargy, dizziness, and loss of muscle coordination. Daily doses of 150 mg interfere with the metabolism of copper and iron [2]. The chapter presents possibility to the lower concentrations of this ion in water, by removing it using bagasse ash of Mozambique red variety sugarcane as adsorbent.

Problem: Environmental pollution caused by the discharge of effluents with contents outside the dumping regulations of Mozambique.

General objective: Characterize and use the bagasse ash of red variety sugarcane, coconut shell, chemically activated carbon, and bentonite of Mozambique as sorbent for zinc removal in model solutions.

2 Materials and Methods

The bagasse ash used was obtained under preset thermal conditions, and the Mozambican coconut shell carbon was obtained by treatment with phosphoric acid 0.6 mol/L and bentonite of Boane region. These materials are characterized from the physical point of view by the determination of pycnometric densities, apparent, apparent by imprisonment, tortuosity, compressibility, and porosity. The determination of the specific surface is possible. In the case of activated carbon, it is necessary to set a series of characteristic parameters for this type of material.

The kinetic of sorption process is evaluated by absorption atomic method. In this study, silicate is brought into contact with known concentration of zinc (II) nitrate solution, for a certain time, to obtain the concentration values of Zn (II) ions at time intervals.

The results obtained from this method are evaluated with the following different kinetic models [1, 3]:

- Pseudo-first order (PFO),
- Pseudo-second order (PSO),
- Elovich model (EM),
- Intraparticle diffusion (DIP).

Related equations are presented below:
Pseudo-first-order model

$$\ln (q_e - q_t) = \ln q_e - k_1 t \quad (1)$$

Pseudo-second order

$$t/q_t = 1/k_2 q_e^2 + t/q_e \quad (2)$$

Elovich model

$$q_t = \alpha + \beta \ln t \quad (3)$$

Intraparticle diffusion model

$$q_t = k_d t^{1/2} \quad (4)$$

where

q_e	quantity of sorbed metal at equilibrium (mg/g)
q_t	quantity of sorbed metal at any time (mg/g)
t	time (min)
k_1	pseudo-first-order velocity constant (min^{-1})
k_2	pseudo-second-order velocity constant (g/mg min)
k_d	intraparticle diffusion velocity constant (mg/g $s^{1/2}$)
α	initial sorption velocity (mg/g min)
β	desorption constant (g/mg)
C_e	concentration of solute at equilibrium (mg/L)
C_o	solute initial concentration (mg/L)
a	constant related to thermal adsorption
k_{ads}	first- and second-order adsorption velocity constant
m	sorbent mass
V	volume of the solution.

3 Analysis of Results

- Bagasse ash of sugarcane characterization
  The bagasse ash of sugarcane used, obtained, has a series of characteristics that are exposed in Table 1.
- Characterization of the coconut shell powder
  The results obtained in the coconut shell powder characterization tests are summarized in Table 2.

Table 1 Sugarcane bagasse ash physic parameters

Parameter	Values
Apparent density	0.62 g/mm^3
Apparent density by imprisonment	0.76 g/mm^3
Pycnometric density	2.3126 g/mm^3
Porosity	73.2%
Compressibility	18.4%
Flow rate	0
Sphericity	0.45
Specific surface	19.863 m^2/g
Tortuosity	1.38

Table 2 Coconut shell powder physic parameters

Parameters	Values
Apparent density	0.15 g/mm^3
Apparent density by imprisonment	0.28 g/mm^3
Pycnometric density	0.418 g/mm^3
Flow rate	0
Tortuosity	1.85
Compressibility	46.64%
Porosity	63.88%
Specific surface	122.32 m^2/g
Relative humidity	8.8%

- Characterization of coal obtained from coconut shell
  Coal obtained from coconut shell has characteristics presented in Table 3.
- Characterization of bentonite
  Below in Table 4 are the results of the bentonite used.
- Kinetic studies
  The results of the kinetic study are shown below in Table 5.

In general terms, the pseudo-second-order model is adjusted to the four adsorbents. However, the Elovich model for bentonite predicts the overlap of a chemical interaction between adsorbate and adsorbent. The adjustment of the diffusional model for bentonite is noteworthy given the high value of the bilinear correlation coefficient obtained.

Below in Table 6 are the values obtained from the parameters corresponding to pseudo-second-order model.

It is worth mentioning the diffusional model in bentonite, where the diffusivity of the zinc (II) ion is 7.52×10^{-14} m^2/s, value within the order for ions that do not have an accentuated polarizability.

Table 3 Coconut coal physical parameters

Parameters	Values
Apparent density	1.07 g/mm^3
Apparent density by imprisonment	1.85 g/mm^3
Pycnometric density	2.726 g/mm^3
Flow rate	0
Tortuosity	0.93
Compressibility	73%
Porosity	49%
Specific surface	219.25 m^2/g
pH	3.5
Volatiles (mf)	22.8
Fixed carbon (mf)	0.6
Ash (mf)	50.6
% Oxygen (maf)	1.5
% Hydrogen (maf)	47.3
% Nitrogen (maf)	1.3

mf: moisture free; maf: moisture and ash free

Table 4 Physic parameters of the bentonite used

Parameters	Values
Apparent density	0.72 g/mm^3
Apparent density by imprisonment	1.05 g/mm^3
Real density	1.8256 g/mm^3
Porosity	60.44%
Compressibility	31.43%
Flow rate	0
Sphericity	0.45
Specific surface	0.056 m^2/g
Tortuosity	1.79

Table 5 Bilinear correlation coefficients for each adsorbent

Model	Ash	Powder	Coal	Bentonite
PFO	0.978	0.952	0.969	0.954
PSO	0.996	0.994	0.995	0.998
EM	0.988	0.979	0.964	0.996
DIP	0.964	0.928	0.919	0.998

Table 6 Parameters of the equation of the pseudo-second-order model

Parameters	Ash	Powder	Coal	Bentonite
h (mg/g min)	14.93	0.004	0.708	2.732
k_2 (g/mg min)	0.51	5.95	3.658	0.35
$t_{1/2}$ (min)	0.36	6.06	0.62	1.02
q_e (mg/g)	5.41	2.34	0.44	2.79

h: Initial velocity
k_2: Apparent constant of reaction rate
$t_{1/2}$: Half-life time
q_e: Adsorptions capacity at equilibrium

4 Conclusions

1. The kinetic model of pseudo-second order is adjusted to the adsorption of the zinc (II) ion in the adsorbents sugarcane bagasse ash, coconut shell powder, chemically activated coconut coal, and bentonite from the Boane region in Mozambique, which advocates a physical-type adsorption.
2. The maximum adsorption capacity of the zinc (II) ions, the maximum initial velocity, and the shortest half-life are represented by the bagasse ash of sugarcane variety Roxa, while coal is the worst of the adsorbents used.

References

1. Chun-I L, Li-Hua W (2008) Rate equations and isotherms for two adsorption models. J Chin Inst Chem Eng 39(6):579–585
2. Dangcong P, Bernet N, Degedenes JP, Moletta R (2000) Effects of oxygen supply methods on the performance of a sequencing batch reactor for high ammonium nitrification. Water Environ Res 72(2):195–200
3. Igwe JC (2006) A bioseparation process for removing heavy metals from waste using biosorbents. Afr J Biotechnol 5(12):1167–1179
4. Metal Environmental Health Criteria 54 (1986) World Health Organization, Geneva, pp 11–15
5. US Environmental Protection Agency (1999) Update of ambient water quality criteria for Zn (II), EPA'822/R-99-014

Diffusivity of Cd (II) Ions in Several Porous Adsorbents

Julio Omar Prieto García, Rafael Quintana Puchol, Alfredo Emilio Curbelo Sánchez, Adrian Alujas Hernández, Joan Rodríguez Díaz, Yennier Cruz Bermúdez and Ángel Mollineda Trujillo

Abstract In this chapter, the diffusivity of the Cd (II) ion in porous adsorbents such as zeolite, clay, coal, and sugarcane bagasse ash is determined. The greater diffusivity responds to the interaction in the heterogeneous solution–coal mass system and the minor to the solution–clay system. Surface and structural factors are the cause of the behaviors described.

Keywords Diffusivity · Cd (II) ions · Porous solids

1 Introduction

Cadmium is an element of the second group in the periodic table. It has a higher polarization capacity than the other elements of this group, which is associated with the greater distortion of the filled layer. This element is very toxic, and it accumulates in the kidneys, blood vessels, and liver with a biological life average in humans from 10 to 30 years. Therefore, it is important to remove from the effluents. Adsorption is one of the methods used for the removal of cadmium in water [3, 5, 6, 8].

J. O. P. García (✉) · R. Q. Puchol · A. A. Hernández · Y. C. Bermúdez
Departamento Licenciatura en Química, Facultad de Química y Farmacia, Universidad Central "Marta Abreu" de las Villas, Carretera a Camajuaní km 5 ½, Santa Clara, Villa Clara, Cuba
e-mail: omarpg@uclv.edu.cu

A. E. C. Sánchez · J. R. Díaz
Departamento de Ingenieria Química, Facultad de Química y Farmacia, Universidad Central "Marta Abreu" de las Villas, Carretera a Camajuaní km 5 ½, Santa Clara, Villa Clara, Cuba

Á. M. Trujillo
Centro de Investigaciones Agropecuarias (CIAP), Facultad de Ciencias Agropecuarias, Universidad Central "Marta Abreu" de las Villas, Carretera a Camajuaní km 5 ½, Santa Clara, Villa Clara, Cuba

R. Cárdenas et al. (eds.), *Proceedings of the 2nd International Conference on BioGeoSciences*, https://doi.org/10.1007/978-3-030-04233-2_13

The diffusivity (also called diffusion coefficient) is the constant of proportionality that appears in Fick's law of diffusion, by relating the density of the molar flow J with the concentration gradient (Δ_C/Δ_X) [1].

$$J = -D(\Delta_C/\Delta_X)$$

Diffusivity is not a property of a substance, since one cannot speak of mass transfer in pure substances. It is a property of a mixture, so when referring to this, it is understood that it is of one substance in another.

Models for estimating diffusion coefficients in liquids are not as accurate as for gases. In a liquid, the diffusivity increases when the temperature increases, decreases when the molecular mass increases, and is almost unaffected by the pressure [7].

The diffusivity depends strongly on the concentration, so in many cases, it can only be estimated for very low concentrations, that is, at infinite dilution (indicated by a zero superscript).

Since the effect of concentration can rarely be estimated, from a practical point of view, it is assumed that diffusivity at infinite dilution is applied for higher concentrations, sometimes as high as 5 or 10 mol% of A in B.

When the concentration of solute increases, the diffusivity deviates from the value to infinite dilution because in addition to the solute–solvent molecular interactions, solute interactions become important.

In heterogeneous systems, it is possible to apply the diffusional model that allows to obtain the effective coefficient and from this, the theoretical diffusivity. It is important to note that the hydrated radius of the ions as well as the polarizing character and the polarizability of them play a prominent role in the diffusive process of the ions in solution through an adsorbent matrix where the active surface, surface area, and dimensions of micropores volume play an important role.

2 Materials and Methods

The following are used as products:

- Natural zeolite from the Tasajeras (Remedios municipality) deposit in Villa Clara Province, Cuba;
- Clay from the deposit located in the municipality of Manicaragua, in the province of Villa Clara, Cuba;
- Sugarcane bagasse ash from the Heriberto Duquesne sugar mill in the municipality of Sagua La Grande Province of Villa Clara, Cuba;
- Coal "chemically" obtained from bamboo.

A characterization of the sorbent is made based on the parameters, apparent density, tortuosity, and surface.

With chemical analysis, materials are characterized from the quantitative point of view, which allows to determine their chemical composition and impurities. The content of silicon, sodium, titanium, phosphorus, calcium, iron, nickel, manganese, copper, antimony, arsenic, thallium, tungsten, lead, magnesium, manganese, cobalt, molybdenum, iron, bismuth, and silver is determined.

The X-ray diffractogram of zeolite is obtained in a Philips model X'Pert diffractometer with $K\alpha$ copper radiation with a voltage of 40 kV and a current of 40 mA.

The equipment used for the determination of the specific surface is BET Gemini III 2375 Surface Area Analyzer adapted to an evacuation pressure of 100 mmHg/min and equilibrium time of 60 s.

Atomic absorption spectrophotometry is used to determine the concentration of Cd (II) ions.

2.1 Kinetics of Adsorption Process

The study is aimed at determining the diffusivity of the Cd (II) ion in several porous adsorbents.

For this, six solutions of 500 mL of 9.8 mg/L concentrations were prepared, and they were placed in contact with 2.0 g of adsorbent, for the time necessary until the concentration remaining in the solution was constant, with constant stirring of 150 rpm using a mechanical agitator. Samples were taken at regular intervals of time, for the determination of the variation of the concentration with respect to contact time. Subsequently, each of the kinetic models used is evaluated, and the constants associated with the models with the best correlation were calculated.

The models evaluated in this work are described below [2, 4, 9]:

Pseudo-zero-order model (SOC)

$$(q_t) = q_e - K_0 t \tag{1}$$

Pseudo-first-order model (SPO)

$$\ln(q_e - q_t) = \ln q_e - K_8 t \tag{2}$$

Pseudo-second-order model (SSO)

$$\frac{t}{q_t} = \frac{1}{k_9 q_e^2} + \left[\frac{1}{q_e}\right] t \tag{3}$$

Elovich model (ME)

$$q_t = \alpha + \beta \ln t \tag{4}$$

Intraparticle diffusion model (DIP)

$$q_t = k_{10}t^{1/2} + C \tag{5}$$

Modelo de difusión en la película de líquido (DPL)

$$\ln(1 - q_e/q_t) = kt \tag{6}$$

Bangham model (MB)

$$\log(C_o/(C_o - q_t m)) = \log\,(k_0 m/2.303V) + \Psi \log t \tag{7}$$

where

V: Adsorption velocity (mg/s L),
C: Concentration of the solution (mg/L),
q: Mass of the adsorbate in the ash (mg/g),
t: Time elapsed since the beginning of the experience (s),
Co: Initial concentration of the solution (mg/L),
q^{+}: Maximum load of the ash (mg/g),
k_0, k_1, …, k_{10}, K, n, K_{ads}, C: They are characteristic constants,
q_e: Adsorption capacity at equilibrium (g/mg),
q_t: Adsorption capacity over time (g/mg),
α: Initial sorption velocity (mg/g min),
β: Sorption constant (g/mg),
m: Mass of the system,
Ψ: Constant.

The mathematical expression that allows to obtain the effective diffusivity from the diffusional model is used:

$$q_t = k_p t^{0.5} \tag{8}$$

where

q_t Adsorption capacity over time (adsorbent mg/adsorbent g),
k_p Diffusional constant (mg/g min$^{0.5}$),
t Time (min).

From this expression, it is possible to obtain effective diffusivity through:

$$k_p = \frac{q_e}{r}\sqrt{\frac{D}{\Pi}} \tag{9}$$

where

q_e Equilibrium adsorption capacity (adsorbent mg/adsorbent g),
k_p Diffusional constant (mg/g min$^{0.5}$),
D Effective diffusivity (cm^2/min),
t Time (min).

When the effective diffusivity is obtained, the theoretical diffusivity is obtained through:

$$D = \frac{D_T d_a}{\Gamma} \tag{10}$$

where

D Effective diffusivity (cm^2/min),
D_T Theoretical diffusivity (cm^2/min),
d_a Apparent density (g/cm^3),
Γ Tortuosity.

Therefore, the apparent density, tortuosity, and specific surface values of the adsorbents used are specified.

3 Analysis of the Results

The values of apparent density, tortuosity, and specific surface of the adsorbents used in the study are shown in Table 1.

Through chemical analysis, the composition of each of the materials used was determined.

Table 2 shows the composition of the zeolite. It is noteworthy that it is a compound rich in Si and Al, as befits an aluminosilicate, with the presence of water lodged in the structure, and a clear predominance of the first element is evident. This composition corresponds to the typical composition of clinoptinolite–heulandite.

Table 1 Values of apparent density and tortuosity of the adsorbents

Materials	Apparent density (g/mL)	Tortuosity	Surface (m^2/g)
Sugarcane bagasse ash	0.34	1.66	11.79
Zeolite	0.87	1.13	178
Clay	0.84	1.16	44.23
Carbon	0.38	1.62	202.15

Table 2 Chemical composition of zeolites

Chemical composition	Content in %	Chemical composition	Content in %
SiO_2	66	CaO	4.50
Al_2O_3	11	MgO	0.50
TiO_2	0.45	Na_2O	2.75
Fe_2O_3	1.75	K_2O	1.00
FeO	0.50	H_2O	11.55

Clay is analyzed after mass distribution, after separation by sedimentation.

- *Clay fraction* Size $\leq$ 0.002 mm **74%** in mass, approximately;
- *Non-clay* fraction Tamaño $\geq$ 0.002 mm **26%** in mass, approximately.

As can be observed, the fraction with a smaller particle size predominates in the clay rock, which is favorable from the point of view of the adsorption capacity, because it increases the specific surface of the system.

The chemical composition for this fraction is reflected in Tables 3 and 4. As in the previous case, it is important to highlight the amount of Si and Al in the structure. An interesting feature is the presence of other metals in considerable quantities such as Fe, Na, Mg, and Ca that are probably housed between the layers of the structure and contribute to the neutralization of the charges on the surface.

The chemical composition for cane bagasse ash was determined by EDS, so data presented are not definitive, but give an idea of the approximate composition, and this is shown in Table 4. The high content of Si is evident, probably due to impurities coming from the mixture of bagasse with soil. There are other important contents such as potassium, calcium, magnesium, phosphorus, chlorine, and sulfur, which do not constitute any irregularity, because as this material comes from the burning of vegetable waste, it is common to find in them different amounts of salts in the form of phosphates, sulfates, chlorides, and soluble salts of calcium and potassium (the latter are very common in vegetable ashes).

From X-ray diffraction, the phases present in the samples are determined by comparison with the corresponding patterns. The degree of crystallinity can be assessed in this way.

The X-ray diffractogram corresponding to zeolitic rock is shown in Fig. 1. It consists of phases that have high crystallinity with predominance of clinoptinolite–heulandite (C–H), anorthite (A), and quartz (Q), although it can be found phases of aluminosilicates of magnesium (S) and modernite (M). This can be favorable for the adsorption capacity since the silicates groups are distributed in three-dimensional form where they originate an intricate network of tunnels that can accommodate different compounds.

The diffractogram of clay fraction is shown below in Fig. 2. There is a predominance of the mineralogical phase corresponding to a montmorillonite (M) of irregular layers; these manifest a high expansive character and ability to exchange ions in their natural state, property that they lose after a thermal treatment at high

Table 3 Chemical composition of clay fraction in clay

Chemical composition	Content in %	Chemical composition	Content in %	Chemical composition	Content in %	Chemical composition	Content in %
SiO_2	57.74	TiO_2	0.58	BaO	0.04	MgO	1.80
Al_2O_3	18.71	Cr_2O_3	0.01	SO_3	0.02	CaO	1.85
Fe_2O_3	7.07	CuO	0.01	P_2O_5	0.07	SrO	0.01
Na_2O	2.68	MnO	0.12	ZrO_2	0.01	WO_3	0.04
K_2O	0.65	NiO	0.01	ZnO	0.01	H_2O	8.58

Table 4 Chemical composition of sugarcane bagasse ash

Chemical composition	Content in %
CaO	6.67
SiO_2	43.55
Cl	1.17
MgO	6.12
K_2O	18.49
Fe_2O_3	0.60
P_2O_5	12.85
MnO	0.48
SO_3	10.05

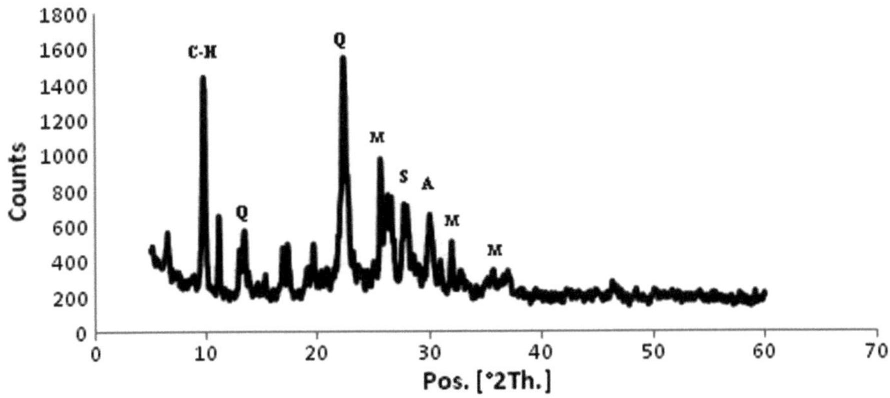

Fig. 1 X-ray diffractogram for natural zeolite

temperatures. Other clayey phases identified are illites (I), which are characterized by being not very expansive and with low ion exchange capacity, in addition to limonite (L) and kaolinite (C). There are also, as accompanying minerals, quartz (Q) and feldspar (F) in the form of albite, these phases, which are present in small quantities, have high crystallinity and do not contribute to the capacity of adsorption, despite its small size.

The case of cane bagasse ash is presented below. The diffractogram in Fig. 3 shows the presence of crystalline phases in the form of cristobalite (Q) in high disorder and feldspathoids (C) which are structures of high crystallinity that do not contribute to the capacity of adsorption; however, the feldspathoid structures are characterized by the presence of channels (24) which would cause some porosity. There is also an amorphous region between 40° and 50° characteristic that is evidenced in the previous studies regarding cane bagasse ash, which should not represent more than 10%, and it is possible that it is due to the lack of uniformity in the burning of the bagasse, which is not done in a controlled manner, and to the

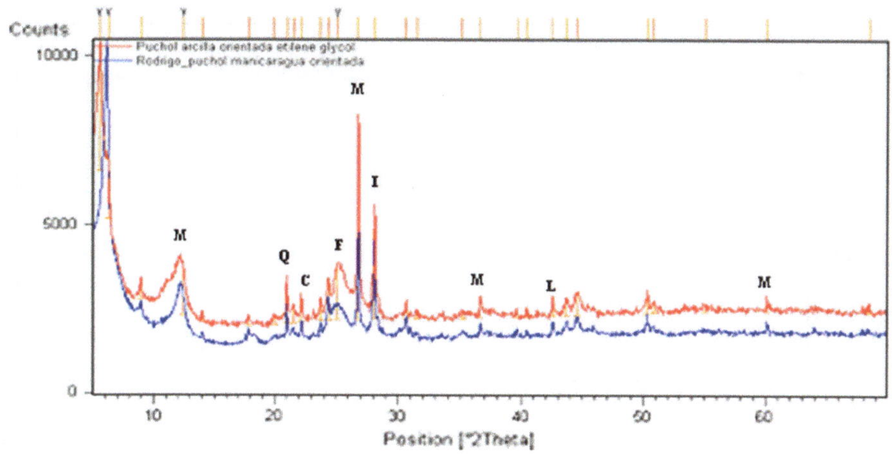

Fig. 2 X-ray diffractogram of the clay fraction

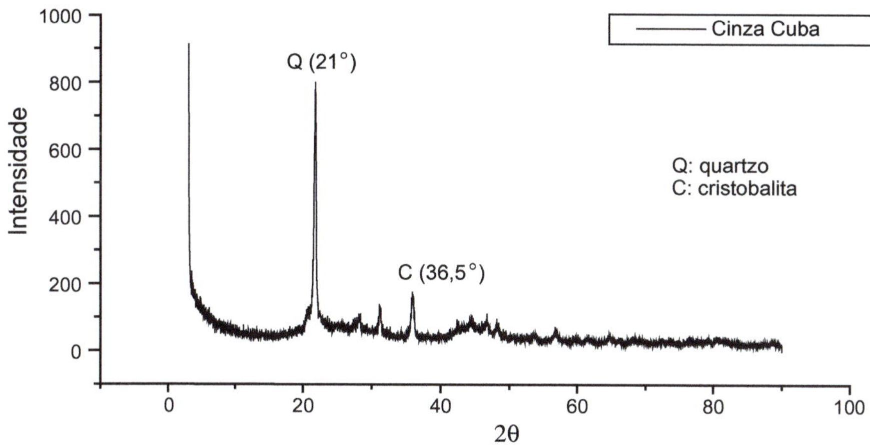

Fig. 3 X-ray diffractogram of sugar cane bagasse ash

region of the oven from which the sample is taken. The amorphous fraction and the present feldspathoid phase could be those with the greatest contribution to the adsorption capacity.

In the diffractogram corresponding to coal, a low crystallinity is observed, with a marked tendency to amorphous phases (Fig. 4).

The ratio of the adsorption capacity of Cd (II) ions (mg/g) over the time (h) for the solid matrices used is established in Fig. 5.

The kinetic study allows to obtain the model that best fits the kinetic requirements (Table 5).

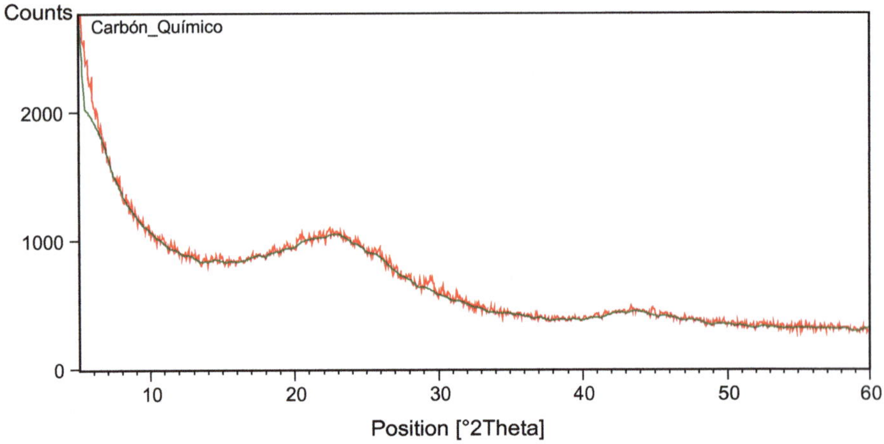

Fig. 4 X-ray diffractogram of physically activated carbon

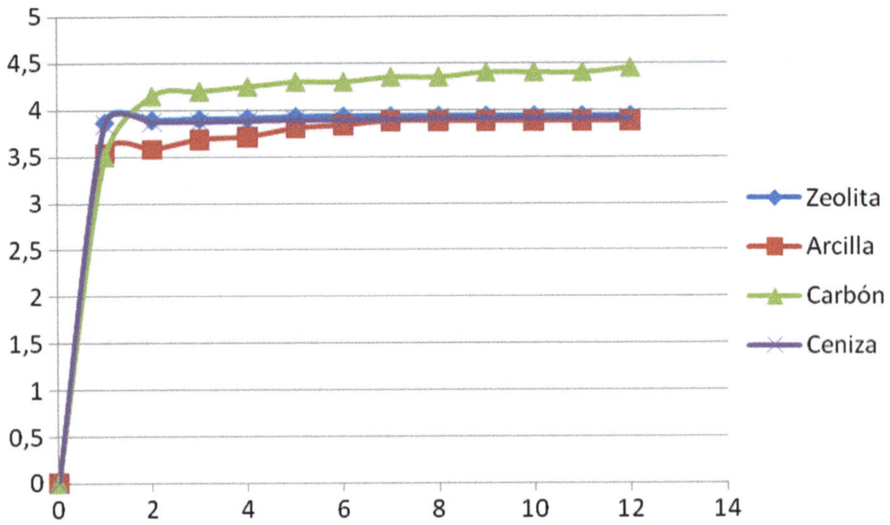

Fig. 5 Adsorption capacity *q* (mg/g) versus *t* (h) of the used adsorbents

It can be seen that the pseudo-second-order model reflects the kinetic process of adsorption, conditioned by a physical adsorption with the interaction of the Cd (II) ion with two points on the surface of the adsorbent. The above is fulfilled in the four adsorbents, independent of their crystalline structure and morphological characteristics. From the diffusive point of view, the mechanism that prevails in bases to the bilinear correlation coefficients is the intercrystalline diffusion.

The higher adsorption capacity of large ions is related to their greater polarizability and lower hydration, which allows them to approach the surface of the solid.

Table 5 Bilinear correlation coefficients of the chosen models

Models	Material			
	zeolite	Clay	Coal	Ash
SOC	0.569	0.481	0.745	0.648
SPO	0.929	0.943	0.909	0.941
SSO	1	0.999	0.999	1
ME	0.950	0.948	0.809	0.932
MB	0.946	0.892	0.8912	0.888
DIP	0.983	0.986	0.984	0.984
DPL	0.911	0.945	0.911	0.956

Table 6 Effective and theoretical diffusivity of Cd (II) in the adsorbents used

Material	Zeolite	Clay	Coal	Ash
Effective diffusivity (m^2/s)	1.42×10^{-14}	4.31×10^{-18}	2.02×0^{-13}	5.93×10^{-15}
Theoretical diffusivity (m^2/s)	1.84×10^{-14}	5.95×10^{-18}	8.61×10^{-13}	1.21×10^{-15}

The relationship of the concentrations in the double layer of ions of equal charge, but having different values of the adsorption potentials $\emptyset_1$ and $\emptyset_2$, is related to the relation of their concentrations in volume by means of the expression:

$$n_1/n_2 = n_{10}/n_{20}\exp[\emptyset_1 - \emptyset_2/kT] \tag{11}$$

which is a particular case of the Nikolski equation. This equation of general form can be expressed by:

$$C_1^{1/Z1}/C_2^{1/Z2} = k_{12}\left[a_1^{1/Z1}/a_2^{1/Z2}\right] \tag{12}$$

Here, a_1 and a_2 are the activities of the corresponding ions in solution and k_{12} the constant of exchange of the corresponding ions determined by the difference of the adsorption potentials. In this case, the "recharge" of the surface can occur related to the variation of the sign of the potential of the adsorption layer linked to the thermodynamic potential of the surface.

From the results of the intraparticle diffusion model or diffusional model, the effective and theoretical diffusivity is obtained (Table 6).

Carbon with a high specific surface and its amorphous structure allows the cadmium (II) ion of high polarizability (0.96×10^{-24} cm^3) to diffuse rapidly in its structure. In zeolite, the aluminosilicatic structure with cavities allows the diffusion of Cd (II) ion in its structure. For ash, a system with crystalline and amorphous phases causes the diffusivity of the ion to be lower than in zeolite and coal. Finally, the crystalline planar structure of clay causes a slow diffusivity of the ion in the structure of the clay matrix.

4 Conclusions

1. The kinetic model that adjusts to adsorption of Cd (II) ions in the zeolite, clay, coal, and sugarcane bagasse ash is of second order;
2. The diffusive model that explains the diffusion phenomenon of cadmium (II) ions in the adsorbents used is that of intraparticle diffusion;
3. The effective diffusion is greater in the coal while it is smaller in the clay in which the amorphous structure of the first and crystalline flat of the second play a fundamental role.

References

1. Benitez J (2009) Principles and modern applications of mass transfer, 2nd edn. Willey, pp 16–17, 41–62
2. Chen H, Wang A (2007) Kinetic and isothermal studies of lead ion adsorption onto palygorskite clay. J Colloid Interf Sci 307:309–316
3. Cotton FA, and Wilkinson G (1988) Química Inorgánica Avanzada, 4th edn. Editorial Limusa, Mexico, p 730
4. Dong L, Zhu Z, Qiu Y, Zhao J (2010) Removal of lead from aqueous solution by hydroxyapatite/magnetite composite adsorbent. Chem Eng J 165:827–834
5. Johnson FM (1998) The genetic effects of environmental lead. Mutat Res 410:123–140
6. Landford PW, Eckenfelder W, Van Nostrand R (1990) Toxicity reduction in industrial effluents. In: A review of potentially low-cost sorbents for heavy metals, pp 98–124
7. Perry RH, Green DW (2004) Manual del Ingeniero Químico, 7th edn. McGraw-Hill, pp 2-328, 2-330, 2-336
8. Saeed A, Iqbal M, Waheed M (2005) Removal and recovery of lead (II) from single and multimetal (Cd, Cu, Ni, Zn) solutions by crop milling waste (black gram husk). J Hazard Mater B117:65–73
9. Tofighy M, Mohammadi T (2011) Adsorption of divalent heavy metal ions from water using carbon nanotube sheets. J Hazard Mater 185:140–147

Increasing Tolerance Plants to Heavy Metals

Evgeny Aleksandrovich Gladkov and Olga Victorovna Gladkova

Abstract Lawn grasses, the important part of a city landscape, contrary to lifeless components, bring the natural color which is a little softening rigidity of surrounding buildings. However, the urban conditions are extremely unfavorable for the growth of lawn grass, because of the high level of contamination with heavy metals. Cell selection technologies have proven themselves in the preparation of plants and cell cultures tolerant to different environmental stressors: drought, low and high temperatures, salinity, but works on obtaining monocots, tolerable to lead and zinc virtually none. The object of our study was lawn grass—*Agrostis stolonifera* L. The authors developed technology for obtaining plants *A. stolonifera*, resistant to lead. Resistant cells were selected after 2–3 subcultivation of calli on modified Murashige and Skoog medium containing 650 mg/l Pb. Regeneration and root formation were performed on Murashige and Skoog medium containing 650 mg/l Pb too. The authors developed technology for obtaining plants *A. stolonifera*, resistant to zinc. Therefore, by means of cell selection, it is possible to raise an ecological valence to lead and zinc and partially to solve the most important environmental problem of city gardening—decorative effect loss, at rather low level of pollution and in certain cases partial degradation of city lawns at the raised level of lead and zinc in a soil cover.

Keywords *Agrostis stolonifera* · Lead · Zinc · Heavy metals

E. A. Gladkov (✉) · O. V. Gladkova
Timiryazev Institute of Plant Physiology, Russian Academy of Sciences, Moscow, Russia
e-mail: gladkovu@mail.ru

R. Cárdenas et al. (eds.), *Proceedings of the 2nd International Conference on BioGeoSciences*, https://doi.org/10.1007/978-3-030-04233-2_14

1 Introduction

Greening of the city is important in the ecology of the urban environment.

Landscaping reduces the level of pollution and the impact of adverse environmental factors affecting urban ecosystems.

Lawn grasses, the important part of a city landscape, contrary to lifeless components, bring the natural color which is a little softening rigidity of surrounding buildings. However, the urban conditions are extremely unfavorable for the growth of lawn grass, because of the high level of contamination with heavy metals.

Heavy metals are priority pollutant soil of cities [1–6]. Sources of entry into the environment of heavy metals:

- vehicle and city transport;
- emissions of industrial enterprises, especially metalworking enterprises and enterprises of ferrous and nonferrous metallurgy, as well as sewage from industrial enterprises;
- waste of housing and communal services;
- heat and power engineering and electric power industry;
- solid waste incineration.

Among the main pollutants of the Moscow soil cover are zinc, lead, copper, and to a lesser extent nickel, cobalt, and cadmium. Thus, among the priority pollutants of the soil are zinc and lead.

In urban ecosystems, lawns have reduced decorative qualities. Among the lawns studied, only 32% had a bald patch of less than 10% [5]. Unfortunately, practically all available lawns are created exclusively for normal ecological conditions; therefore, creation of lawns in conditions of a high level of contamination is a difficult task.

Cell selection technologies have proven themselves in the preparation of plants and cell cultures tolerant to different environmental stressors: drought, low and high temperatures, salinity, but works on obtaining monocots, tolerable to lead and zinc virtually none [7–9].

2 Objects and Methods

The object of our study was a perennial grass *A. stolonifera* L. This lawn grass is widely used in landscaping for lawns for various purposes. The degree of phytotoxicity was assessed by shoot growth of *A. stolonifera* L. on filter paper placed in petri dishes, as well as by sowing in soil (pH 6.5–7.0, nitrogen, phosphorus, and potassium contents, respectively, 150, 250, and 300 mg/l) containing zinc and lead. The control plants grow in aqueous solution or in soil without zinc and lead.

To select tolerant clones, calli were cultivated on Murashige and Skoog medium with 1 mg/l 2, 4-D, and 650 mg/l Pb $(NO_3)_2$ (data are presented in terms of pure metal). Regeneration and root formation were performed on Murashige and Skoog medium containing 650 mg/l Pb $(NO_3)_2$. To select tolerant clones, calli were cultivated on Murashige and Skoog medium with 1 mg/l 2, 4-D, and 300 mg/l zinc sulfate $ZnSO_4*7H_2O$ (data are presented in terms of pure metal). Regeneration and root formation were performed on Murashige and Skoog medium containing 300 mg/l zinc sulfate.

3 Results and Discussion

Experiments in aqueous solutions of lead showed phytotoxicity to the plants at relatively high concentrations. At a concentration of 650 mg/l of lead, inhibition growth of shoots was about 30% of the control. A slight decrease in the growth of shoots of *A. stolonifera* in soil conditions was observed at 650 mg/kg; this concentration is comparable to the highest level of soil contamination in Moscow.

To select tolerant plants, the methods of introduction into the culture in vitro, culturing, and plant regeneration were developed.

The addition of 650 mg/l of lead caused a reduction of tissue growth in two times (Fig. 1) the part of the calli died on the medium with lead, and some maintained a growth level of control. Lethal concentration of lead is 1600 mg/l.

With regard to effect of Pb on shoot regeneration the frequency of calluses with shoots decreased significantly with increasing Pb concentration starting at 650 mg/l Pb (Table 1). At 650 mg/l regeneration was 50.5%. At 1200 mg/l regeneration ability was completely inhibited.

The concentration of 650 mg/l Pb was chosen as selective for calli *A. stolonifera.*

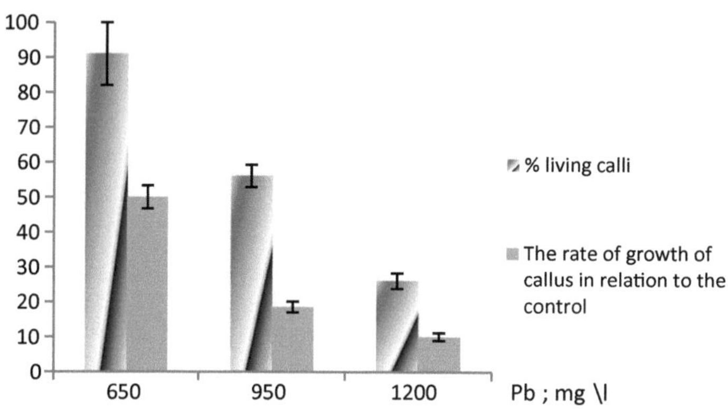

Fig. 1 Effect of lead on the growth and death of calli of *Agrostis stolonifera* L.

Table 1 Effect of lead on the regeneration ability of calli of *Agrostis stolonifera* L

Pb; mg/l	% Regeneration
650	50.5 ± 4.1
950	33.0 ± 2.8
1200	0

The authors developed technology for obtaining plants *A. stolonifera*, resistant to lead.

The selective factor was presented in the medium at all stages of selection, including regeneration and rooting of shoots, in order to increase the probability of producing of resistant plants. Calli were incubated in petri dishes on filter paper moistened with a solution of lead (Fig. 2). Resistant cells were selected after 2–3 subcultivation of calli on modified Murashige and Skoog medium containing 650 mg/l Pb. Regeneration and root formation were performed on Murashige and Skoog medium containing 650 mg/l Pb too (Fig. 3).

Most of the test plants produced from lead-resistant cells were more tolerant to lead than original plants. Test plants fully retain high decorative quality when planted in soil containing lead (Fig. 4).

Experiments in aqueous solutions of zinc showed phytotoxicity to the plants at relatively low concentrations.

At a concentration of 80 mg/l of zinc, inhibition growth of shoots was 27% of the control (shoot growth 73% of control) (Fig. 5).

At a concentration of 100 mg/l of zinc observed a marked slowdown in the growth of shoots (inhibition growth of shoots was 33% of the control).

The results obtained in soil conditions were associated with aqueous solutions (Table 2). Plants preserved decorative qualities at concentrations of zinc in soil less permissible concentration [10].

The toxicity of zinc for callus cultures was determined. The concentration of 300 mg/l Zn was chosen as selective for calli (Table 3).

Fig. 2 Calli *Agrostis stolonifera* cultivated on Murashige and Skoog medium with lead

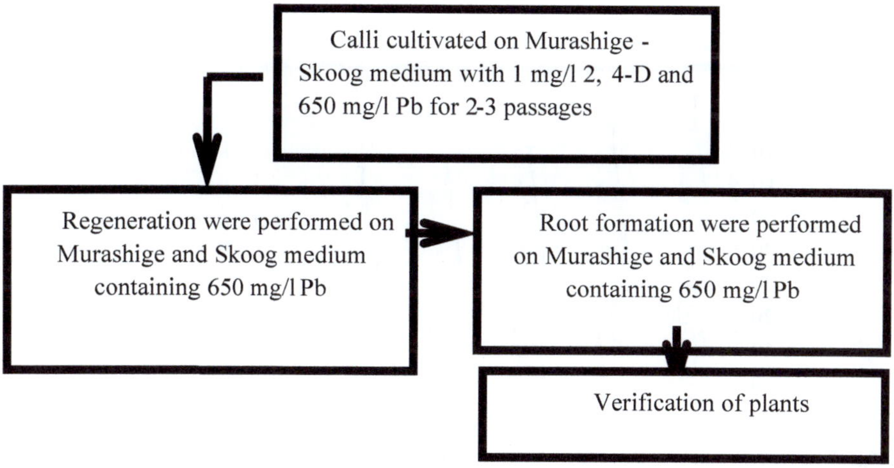

Fig. 3 Cell selection scheme for producing plants tolerant to lead

Fig. 4 Plants *Agrostis stolonifera*, after cell selection

Calli were incubated in petri dishes on filter paper moistened with a solution of zinc. The selective factor was presented in the medium at all stages of selection, including regeneration and rooting of shoots, in order to increase the probability of producing of resistant plants. Resistant cells were selected after 2–3 subcultivation of calli on modified Murashige and Skoog medium containing 300 mg/l Zn. Regeneration and root formation were performed on Murashige and Skoog medium containing 300 mg/l Zn too (Fig. 6).

To test the sustainability of regenerants obtained after cell selection. Regenerants and control plants were planted in soil containing 450 mg/kg Zn. The most of the tested plants produced from zinc—resistant cells were more tolerant to zinc than control plants.

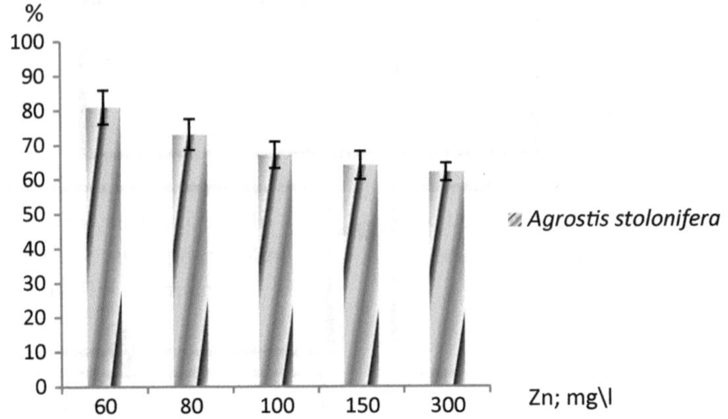

Fig. 5 Effect of zinc on shoot growth of ornamental plants in aqueous solutions (% relative to control)

Table 2 Effect of zinc on shoot growth of ornamental plants in soil (% relative to control)

Zn; mg/kg	*Agrostis stolonifera*
40	91 ± 9.6
60	78 ± 8.8
80	75 ± 10.0
100	64 ± 7.9

Table 3 Effect of zinc on the growth and death of calli of *Agrostis stolonifera* L.

Zn; mg/l	% Living calli	The rate of growth of callus in relation to the control
300	48 ± 4.8	50 ± 3.2
350	33 ± 3.6	30.4 ± 3.0
450	21 ± 2.2	18.6 ± 1.8
600	5 ± 0.5	0

Thus, zinc showed greater phytotoxicity than lead. Zinc is the limiting factor for plants *A. stolonifera*. Our experiments have been shown negative effects on plants, even at low concentrations. Assessment of soil pollution based on hygienic standards cannot guarantee the absence of effects on plants.

Thus, technologies for obtaining *A. stolonifera* plants resistant to lead and zinc have been developed. Therefore, by means of cell selection, it is possible to raise an ecological valence to lead and zinc and partially to solve the most important environmental problem of city gardening—decorative effect loss, at rather low level of pollution and in certain cases partial degradation of city lawns at the raised level of lead and zinc in a soil cover.

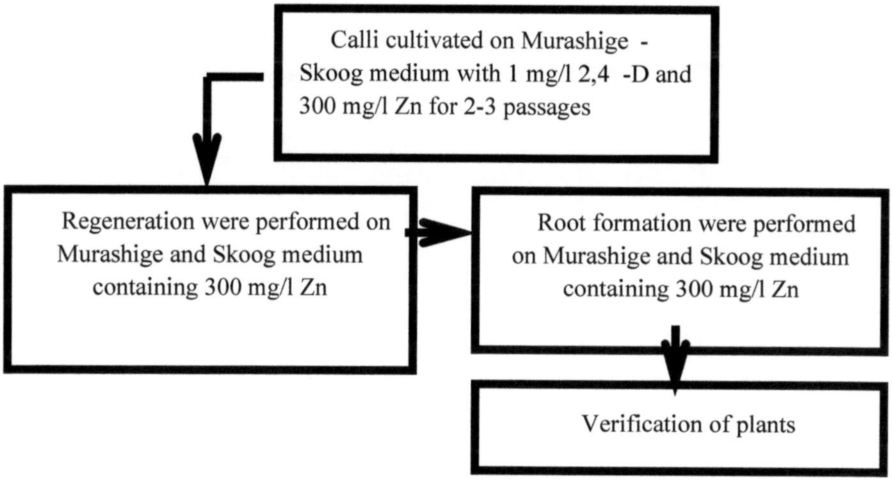

Fig. 6 Cell selection scheme for producing plants tolerant to Zn

References

1. Gladkov EA, Dolgikh YI, Gladkova OV (2012) Phytotechnology for environmental protection: textbook. M.: Moscow State University of Engineering Ecology, pp 202
2. Gladkov EA, Dolgikh YI, Gladkova OV (2016) Assessing the possibility of the use of cell selection in phytoremediation. Asian J Microbiol, Biotechnol Environ Sci, Glob Public. 18 (2):499–502
3. Gladkov EA, Gladkova OV (2007) A method for producing tolerance to the combined effect of heavy metals monocotyledonous plants in vitro. Patent No 2311021. 33
4. Report on environment condition of Moscow city in 2010 (2011). Department of Natural Resources and Environmental Protection of the City of Moscow. Moscow
5. Report on environment condition of Moscow city in 2013 (2014). Department of Natural Resources and Environmental Protection of the City of Moscow. Moscow
6. Report on environment condition of Moscow city in 2016 (2017). Department of Natural Resources and Environmental Protection of the City of Moscow. Moscow
7. Gladkov EA, Dolgikh YuI, Gladkova OV (2014) In vitro selection for tolerance to soil chloride salinization in perennial grasses. Sel'skokhozyaistvennaya Biologiya (Agric Biol) 4:106–111
8. Gori P, Schiff S (1998) Response of in vitro cultures of *Nicotiana Tabacum* L to copper stress and selection of plants from Cu—tolerant callus. Plant Cell, Tissue Organ Cult 53(3):161–169
9. Vera-Estrella R, Miranda-Vergara MC, Barkla BJ (2009) Zinc tolerance and accumulation in stable cell suspension cultures and in vitro regenerated plants of the emerging model plant *Arabidopsis halleri* (Brassicaceae). Planta 229(4): 977–986
10. Hygienic standards as of 01.01.1991 (1990), The State Committee of the USSR, 02–2333, 10. 12

Increasing Tolerance *Agrostis Stolonifera*, *Festuca Rubra*, *Brachycome Iberidifolia*, *Chrysanthemum Carinatum* to Copper

Evgen Aleksandrovich Gladkov, Ilina Igorevna Tashlieva, Yuliya Ivanovna Dolgikh and Olga Victorovna Gladkova

Abstract There were used biotechnological methods for stress-tolerant lines through tissue culture based in vitro selection. The authors developed technologies for obtaining plants *Agrostis stolonifera* L., *Festuca rubra rubra*, *Brachycome iberidifolia Benth.*, *Chrysanthemum carinatum Schousb.* resistant to copper. It was carried on a previously investigation that these plants are very sensitive to copper. For each object of the study was determined a selective concentration of copper and developed selection scheme. The scheme selection of *A. stolonifera* and *F. rubra rubra* included two–three subcultivations on the medium for callus growth, one passage on the medium for shoot regeneration, and one passage on the medium for rooting. All media contained 150 mg/l copper. The scheme selection of *B. iberidifolia* and *C. carinatum* consists of culturing callus on medium with addition a concentration of copper 20 mg/l during two passages, regeneration, and rooting on mediums without toxicant.

Keywords *Agrostis stolonifera* · *Brachycome iberidifolia* · *Chrysanthemum carinatum* · Copper · Lawn grasses · Biotechnology

1 Introduction

The plants are an important part of the urban ecosystem. Lawn grasses are a basis of a grassy cover of Moscow, their role in the conditions of an adverse ecological situation continuously grows. Purpose and environmental conditions determine the choice of plants for the lawn. For example, *Agrostis stolonifera* L. forms the highest quality lawn. In addition to traditional lawns, meadow grasses are popular too. Meadow grasses consist of annual flowering grasses, for example, *Brachycome*

E. A. Gladkov (✉) · I. I. Tashlieva · Y. I. Dolgikh · O. V. Gladkova
Timiryazev Institute of Plant Physiology, Russian Academy of Sciences, Moscow, Russia
e-mail: gladkovu@mail.ru

Y. I. Dolgikh
All-Russia Research Institute of Agricultural Biotechnology, Moscow, Russia

R. Cárdenas et al. (eds.), *Proceedings of the 2nd International Conference on BioGeoSciences*, https://doi.org/10.1007/978-3-030-04233-2_15

iberidifolia Benth., *Chrysanthemum carinatum* L. Unfortunately, these plants as well as most ornamental plants are very sensitive to pollution of urban soil.

Among the most dangerous pollutants for plants are copper ions [1]. Copper is one of the main pollutants of urban soil, and it can lead to inhibition of plant grown, dark leaves, root elongation, height shoots reduction [2]. In direct contact with high concentrations of copper in the above-ground plant organs observed twisting and falling of the leaves. Young tissues and organs are particularly susceptible to excess copper. Damaged landscape can be restored with the help of biotechnology. Biotechnological methods allow keeping the biodiversity of grass cover.

Cellular selection is environmentally safe technology to create sustainable forms of plants. For example, cell selection was used to produce resistant plants and cell lines to heavy metals and chloride salinity [3–5]. However, there is almost no work of improving the sustainability of urban plants to copper.

2 Objects and Methods

2.1 Description of research objects

The objects of our study were grasses and flowering plants. *A. stolonifera* L. is a perennial grass, it has an advantage over many other lawn grasses. The lawn from *A. stolonifera* does not need to be cut often, and it withstands a shadowing and relatively resistant to gases (Fig. 1).

Fig. 1 *A. stolonifera*

B. iberidifolia Benth. is an annual herb, Asteraceae. The leaves of *B. iberidifolia* are fully divided, each having long and narrow segments from the midrib. The ray florets are varied in color, white through pink and blue to violet (Fig. 2).

B. iberidifolia used for decoration of the meadow grasses, flower beds, curbs, rocky hills.

Festuca rubra rubra is a perennial grass and is widely used in all types of lawns. Dark green glossy leaves fescue remain even during drought (Fig. 3). *F. rubra* persists in herbage 10 and more years.

C. carinatum Schousb. is annual plant, Asteraceae, height up to 60 cm. This plant is widely used in urban landscaping for growing in groups, mixborders, near shrubs (Fig. 4).

Fig. 2 *B. iberidifolia*

Fig. 3 *F. rubra*

Fig. 4 *C. carinatum*

2.2 In Vitro Selection

For getting callus it was used seeds. The methods for obtaining calluses, regeneration, and rooting of all plants were developed previously [6, 7]. Cultures were incubated in petri dishes on filter paper moistened with a solution of copper at 26 °C, exposed to a 16-h photoperiod in the light and humidity 70%. Callus cultures of *A. stolonifera*, *F. rubra*, *C. carinatum* were inoculated on Murashige and Skoog (MS) medium [8] supplemented with 2, 4-diclorophenoxiacetic acid (2, 4-D), 3% sucrose, 0.05% mg/l caseinhydrolysat, 0.7% agar-agar. Then callus was transferred to MS medium for regeneration and rooting. Callus of *B. iberidifolia* was cultured on Gambourg medium [9] supplement with 3% sucrose, 2 mg/l 2, 4-D, and 2 mg/l kinetin. Then embryo callus was transferred to ½MC medium supplement with 1 mg/l 6-Benzylaminopurine (BAP) and 0.1 mg/l 1-Naphthaleneacetic acid (NAA). Callus of *C. carinatum* was cultured on MS medium supplement with 3% sucrose, 1 mg/l BAP, and 0.1 mg/l Indole-3-acetic acid. For regeneration, callus was transferred to ½MC supplement with 0.5 mg/l BAP and for rooting −½MC supplement with 1% sucrose, 0.1 mg/l NAA.

Concentration of selecting agents for *A. stolonifera and F. rubra* was 150 mg/l $CuSO_4 * 5H_2O$. Callus of this plants was cultured on MS medium with addition 150 mg/l $CuSO_4 * 5H_2O$. After culturing for 1-month light callus with increased in size was collected. Then calli were transplanted on regeneration and rooting medium supplemented with copper at a concentration 150 mg/l.

Concentration of selecting agents for *B. iberidifolia* and *C. carinatum* was 20 mg/l $CuSO_4 * 5H_2O$. Calli were cultured during 26 days and then induction shoot and root on the mediums without copper.

3 Results and Discussion

For in vitro selection, it was chosen shock treatment, in which cultures are directly subjected to a shock of high concentration and only those which would tolerate that level will survive [10] (Fig. 5). The selective factor was present in the mediums at all stages of selection, including regeneration and rooting of shoots, in order to increase the probability of producing of resistant plants.

The authors developed technology for obtaining plants *A. stolonifera*, resistant to copper. The selection scheme included two–three subcultivations on the medium for callus growth, one passage on the medium for shoot regeneration and one passage on the medium for rooting (Fig. 6). All media contained 150 mg/l copper.

An increase of culturing time up to four months results in decrease of regenerative capacity. 78 out of 180 regenerated *A. stolonifera* plants were successfully rooted in the soil. To test the resistance to high concentrations of copper, thirty regenerants after cell selection were planted in soil containing 150 mg/kg copper (Fig. 7). The most of the tested plants produced from copper-resistant cells were more tolerant to copper than original plants.

The developed approaches for obtaining plants resistant to copper were used for *F. rubra.*

Calluses of *B. iberidifolia and C. carinatum* were much more sensitive to copper then cell tissue of *A. stolonifera* and *F. rubra*. We have been shown previously that

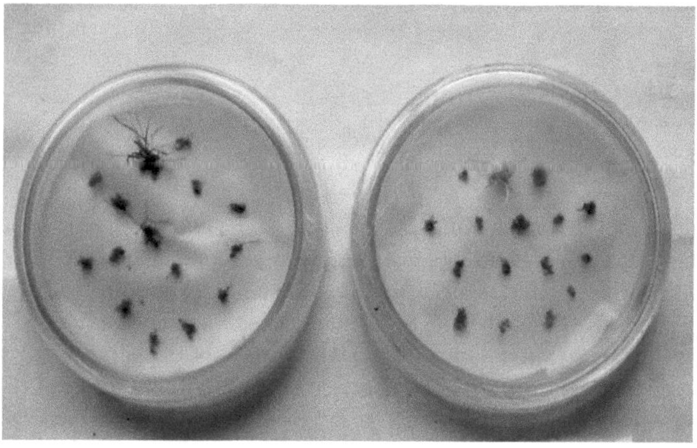

Fig. 5 Plant regeneration of *A. stolonifera* on the medium with copper

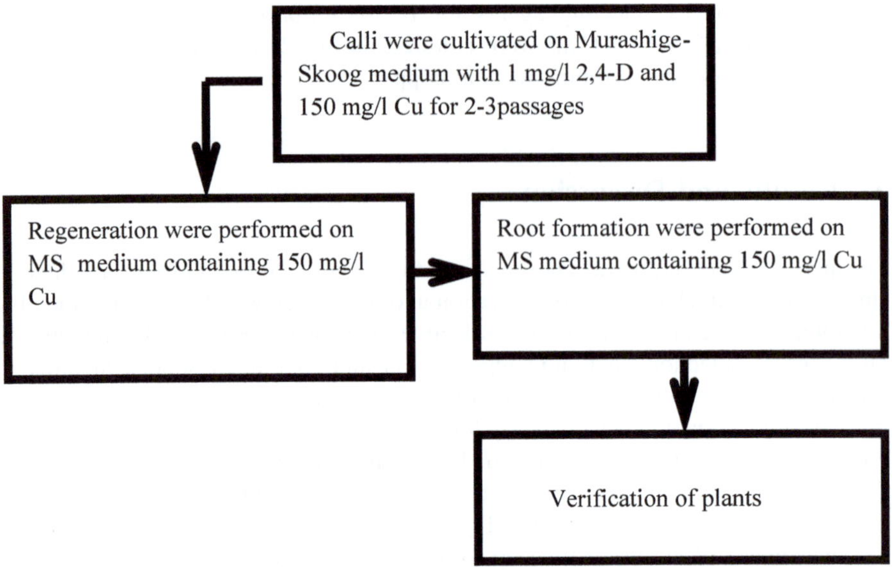

Fig. 6 Cell selection scheme of *A. stolonifera* for producing plants tolerant to copper

Fig. 7 Reentrants of *A. stolonifera* after cell selection

callus cultures of *B. iberidifolia* and *C. carinatum* were tolerant to 10 mg/l of copper, but at the concentration of 20–30 mg/l copper toxic effect is greatly enhanced. Therefore, it was necessary to substantially modify the approach for obtaining plants resistant to copper. It was chosen scheme selection consists of culturing callus on MS medium with a concentration of copper 20 mg/l within two passages, regeneration for 3–4 passages and rooting without toxicant (Fig. 8).

As a result of the selection were received 27 plants of *B. iberidifolia*. Most obtained regenerants possessed increased resistance to 20 mg/l of copper.

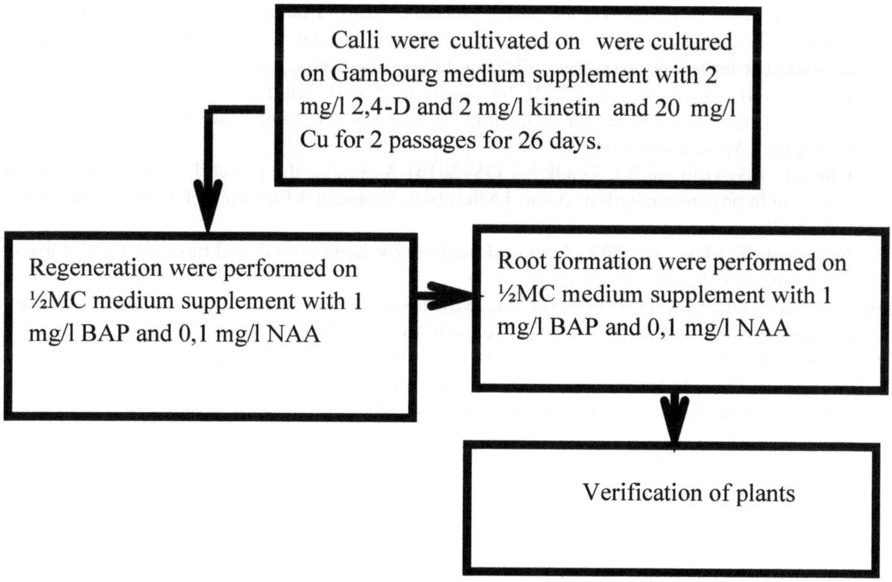

Fig. 8 Cell selection scheme of *B. iberidifolia* for producing plants tolerant to copper

The developed approaches for obtaining plants resistant to copper were used for *C. carinatum*. As a result of the selection was received 21 plants of *C. carinatum*, 10 plants have rooted.

Thus, there were 2 concepts for obtaining resistant plants with in vitro selection. In the first case, the method was suitable for more resistant plants (*A. stolonifera, F. rubra)* to copper. Cell selection and regeneration were carried out on media with a toxicant. On the other hand, *B. iberidifolia* and *C. carinatum* were very sensitive to copper ions, so the cultivation of calli was carried out only in the first stage, then regeneration and rooting were made without copper.

References

1. Gladkov EA, Dolgikh Yu I, Gladkova OV (2012) Phytotechnology for environmental protection: textbook. M: Moscow State University of Engineering Ecology. pp 202
2. Kabata-Pendias A, Pendias X (1989) Copper. In: Trace elements in soils and plants, Moscow, pp 118
3. Gladkov EA, Dolgikh YuI, Gladkova OV (2014) In vitro selection for tolerance to soil chloride salinization in perennial grasses. Sel'skokhozyaistvennaya Biologiya (Agricu Biol). 4:106–111
4. Gori P, Schiff S (1998) Response of in vitro cultures of *Nicotiana Tabacum* L to copper stress and selection of plants from Cu—tolerant callus. Plant Cell, Tissue Organ Cult 53(3):161–169

5. Vera-Estrella R, Miranda-Vergara MC, Barkla BJ (2009) Zinc tolerance and accumulation in stable cell suspension cultures and in vitro regenerated plants of the emerging model plant *Arabidopsis halleri* (Brassicaceae). Planta 229(4):977–986
6. Litvinova II, Gladkov EA (2012) Introduction to cell culture in plants used as fodder, medicinal and decorative for obtaining stress-resistant forms. Sel'skokhozyaistvennaya Biologiya (Agric Biol) 4:94–99
7. Gladkov EA, Dolgikh YI, Gladkova OV(2016) Assessing the possibility of the use of cell selection in phytoremediation. Asian J Microbiol, Biotechnol Environ Sci, Glob Public. 18(2): 499–502
8. Murashige T, Skoog F (1962) A revised medium for rapid growth and bioassaya with tobacco tissue cultures. Physiol Plant 15:473–476
9. Gambourg OL, Elevegh D (1968) Culture methods and detection of glucanases in suspension cultures of weat and parleys. Can J Biochem 46:417–421
10. Purohit M, Srivastava S, Srivastava PS (1998) Stress tolerant plants through tissue culture. In: Srivastava PS (ed) Plant tissue culture and molecular biology: application and prospects. Narosa Publishing House, New Delhi, pp 554–578

Environmentally Sustainable Management of Rice Cultivating Zones in Ukraine

Vasyl Petrenko

Abstract More than half the world's population depends on rice, which is grown on nearly 140–150 million ha of land for a global production of more than 520 million tons annually. As for Ukraine, it is quite new introduced crop which has great potential in Southern Ukraine. But cultivating rice in Danube and Dnepr, lots of problems remain unresolved: the danger of secondary soil salinization, the deficit of irrigation water, there is no drainage water utilization system, water and wind erosion bring significant losses, imperfect technologies of crop protection, there still might occur significant technological discharges containing residues of herbicides, pesticides, and mineral fertilizers into the basins of the Black and Azov Seas. In the article below, we focus on capabilities that maintain soil fertility, preserve or even improve the rice-growing environment in Ukraine, and provide possible opportunities for sustainable land usage beyond rice monoculture in Danube and Dnepr deltas to avoid some adverse effects listed above.

Keywords Paddy field · Irrigation system · Ukraine · Rice cultivating Environment

1 Introduction

It is well known that rice (*Oryza sativa L.*) fields are integral part of the landscape throughout most of the subtropics and tropics. Rice is a major crop that is planted in the most diverse environmental conditions; from sea level, to 3000 m altitude in the Himalayans, from 35° S in Australia to 49° N in the Czech Republic. Temperatures and day lengths are therefore quite diverse in the rice-growing areas. Rice is also grown under different water conditions, from mesoxeric conditions to deepwater reaching up to 5 m [1–3].

V. Petrenko (✉)
Institute of Food Resources, NAAS of Ukraine,
4a Sverstyuk Str., Kyiv 02660, Ukraine
e-mail: whippet85@mail.ru

R. Cárdenas et al. (eds.), *Proceedings of the 2nd International Conference on BioGeoSciences*, https://doi.org/10.1007/978-3-030-04233-2_16

In the world, Ukraine is always associated with large cereal fields and usually is mentioned like European breadbasket. Nowadays, Ukraine is one of the largest producers of corn, wheat, barley, and sunflower seeds not only in Europe but in the world too [4, 5]. Despite the fact that Ukraine is located in temporary climate zone, it has significant areas with rice exclusively japonica type (*sino-japonica*).

In the past when Ukraine was the part of the Russian Empire and then the USSR, rice appeared in our country only in the beginning of the twentieth century. So, it was unrepresentative cereal as distinct from wheat, rye, and barley. At first in the USSR, rice had been grown in the Middle Asia and the South Caucasus areas but with developing of cotton production in those regions, rice fields curtailed drastically. By the fact that rice needs much water then cotton and water is really limiting factor in this region. That is why weather conditions and absoluteness of water resources allowed transferring rice plantings to the southern part of Ukraine, mainly to Danube and Dnepr deltas [6].

2 Engineering Nonconformance of Irrigation Systems

Rice cultivation systems can be considered natural-engineering or natural-anthropogenic systems as they consist of two components: (1) natural, that is determined by natural factors (climate, soil, groundwater, flora, microorganisms, etc.) and (2) anthropogenic, which is attributable to human economic activities (construction of channels, roads, and field levees; planning of paddy fields arrangement that involve some tilth-top soil hauling; fertilizer and herbicides treatment, and most crucially supply of water for irrigation entailing the change of original air and water regime of the particular territory) [7, 8]. Talking about former USSR countries, it is obvious that the design of rice irrigation system did not consider the environmental conditions formed on the basis of natural steppe ecosystems exposed to insufficient atmospheric moistening, automorphic, and semi-hydrogenic soil regime and is dramatically different compared to the growing conditions of the rice culture, requiring long-term paddy field flooding. In view of this, paddy fields should be considered rather temporarily marshy than steppe ecosystems.

At the time of designing the rice cultivation systems in 60–70th of the XX century, there was not enough knowledge and experience in Ukraine to make correct estimates of the influence paddy culture would produce on the environment. It was generally believed that washing out salts would be enough to increase soil fertility; however, the influence of rice irrigation systems on the state of the agricultural landscape of the Black Sea region appeared to be much more complicated [9].

On the territory of Ukraine, most of the irrigation channels feeding paddy fields were constructed just in the ground bed, only some being covered with concrete slabs and only two constructed with the use of concrete ducts. The channels of drainage-waste system were totally built in the ground bed; some channels being strengthened with reinforced concrete slabs. Most of the plots are of the Krasnodar type and consist of paddy fields of rectangular shape. That was a reason which caused appearance and then establishing of different sedge species and some other native wetland plants through the rice cultivating areas. The paddy fields are arranged across the plot and extend from the irrigator to the spillover. Water supply of every paddy field is supported with the plot irrigation channel, whereas water removal is conducted via the plot outlet channel [10–12].

It is necessary to notice that the strategies for mitigating salinity problems in rice production include both the development of management options and the genetic improvement of salinity tolerance in current cultivars. Although the use of some management options can ameliorate yield reduction under salinity stress, their implementation is often limited because of cost and availability of good quality water resources [11, 13, 14]. Therefore, the need for genetic improvement of salt tolerance is great and is expected to increase dramatically in the nearest future.

The existing schemes used for water management of paddy fields are characterized by huge water loses due to technical imperfections of their constructions, absence of proper irrigation rotation, poor rate of drainage/wastewater reuse, the presence of technically unsound and unproductive water removal. All these result in the increase of irrigation rate up to 25–30 thousand m^3/ha while for the most paddy culture lands of the south of Ukraine, characterized by heavy and mediate soil texture and the groundwater level of 1–3 m, the net irrigation rate ranges from 14 thousand m^3/ha to 20 thousand m^3/ha [15].

The Black Sea and Sivash regions favoring agroclimatic conditions for rice growing exist over approximately 1 million ha of land of which around 200 thousand ha are low-output soils. The area which is technically suitable for paddy culture in Ukraine comprises more than 300 thousand ha (Fig. 1), of which in Kherson Oblast—172, in Odessa Oblast—37, in Mykolaiv Oblast—34, in Zaporizhzhya Oblast—65, in Zakarpattia Oblast—12 thousand ha [16].

Growing rice in Ukraine and some other Eastern Europe countries is carried out exclusively with the use of rice irrigation systems which represent complex irrigation utilities designed for supplying water in the field, providing and maintaining the necessary level of water in it throughout the vegetation season, as well as quick water removal from the paddy fields and fall in groundwater level for the rice harvesting period and in winter. In the late 80s, the development of irrigation and drainage in the southern region of Ukraine resulted in the increase of drainage and wastewater outlet to about 2 km^3, and the application of chemical crop-protecting agents, chemical ameliorants, and mineral fertilizers created the severe environmental conditions in the Black and Azov Seas basins. The most unfavorable influence was due to the rice irrigation systems. For example, about 0.65 million tones or 37% of the 2 million tones drainage and wastewater outlet discharged each day to the sea basins was due to the rice irrigation systems, whereas they comprise only 4% of the irrigated land of the south of Ukraine. Paddy culture area of Ukraine spreads over the territory encompassing a considerable amount of recreational resources. In the late 80s of the XX century, in the southern region of Ukraine, there arose significant contradictions: either rice or recreation area. It has brought about

Fig. 1 Rice cultivating areas in Ukraine (marked in green)

hasty decisions concerning prohibition of paddy culture development. Rice has been declared to be the death-threatening product and rice grower to be the concealers of a crime [17–19].

Significant water discharges from rice irrigation systems as well as industrial discharges and wastewater negatively affected the hydrological and ecological regime of the territory and the marine area of the Black and Azov Seas. There is a loss of healing properties of the sea water and a sharp decrease in its mineralization in the coastal zone. In the early 90s of the XX century, biological productivity as well as opportunities for the self-clarification of the Black Sea decreased significantly, and consequently, a significant part of the Black and Azov Seas coasts rendered unsuitable for recreational purposes. Although not all the ecological problems of that time were associated with paddy culture industry development, most of those were attributed to this industry. In this connection "green" environmental groups stroke the active campaign against paddy culture industry which came to tough contradictions between the public and rice growers. The situation was complicated by the mass media, who published articles about the negative impact of the rice industry on the ecological state of the southern region of Ukraine. Owing to this situation since the 90s of the XX century, the area of rice systems began to decline and rice was gradually removed from crop rotation on the rice-growing area of Ukraine. However, these measures did not significantly improve the environmental situation in the region. Truth to be told the same situation was observed in some other countries of the Black Sea basin [20–22].

Another reason of reducing rice paddy areas in Ukraine is the shortage of warmth and sunny days. Temperature affects not only the growth but also the development of the rice plant. Usually effective heat sum has been within 3000–3600 °C, and frost-free period consists of 155–210 days. Accumulated, it allows growing only early ripening and some mid-duration rice varieties. But in some seasons with poor weather conditions, rice productivity in Ukraine was above 3 tones per ha and even less (generally the rice yield is about 5.0–5.2 tones per hectare). Low temperature during the reproductive and ripening phase of the rice plant usually results in high shattering percentage of the grains in several of the high-yielding modern rice varieties [23–25].

3 Regional Particularities

Each region is characterized by its specific environmental conditions; that means, each region represents the super complicated ecological system that includes such subsystems as nature, society, and production. Being interconnected and interdependent, they should develop in such a way that would support their complemental coexistence on the basis of conflict-free usage of natural resources potential. The point at issue is about providing the adequacy of interests for preservation of the environment, solution of contradictions between development courses of different branches of economy in the region, and timely addressing conflicts of interests in the field of natural resources management [26].

The natural resources potential of the Black Sea region promotes the development of various economic sectors of the region—industry, agrarian sector including paddy culture, and recreational and touristic complex. The agrarian potential of the paddy culture region of Ukraine is quite significant due to its land and aquatic resources, natural and climatic conditions, etc. The availability of natural resources by itself does not support, but in many cases rather impedes economic development, stimulating a rent-seeking economic behavior. On the contrary, the regions with poor natural and climatic conditions demonstrate activity and inventive power concerning implementation of sustainable technology and innovative solutions. The agrarian sector of the economy is one of the most powerful business entities in the Black Sea region. It is represented by the intensive irrigation agriculture over the area of more than 1 million ha in Ukraine only [27, 28].

There were lots of reasons for contradiction. Growing crops, including rice, with intensive technologies, application of significant amounts of herbicides, pesticides, and fertilizers, usage of unrestricted volumes of water for irrigation, as in the Soviet era its cost was negligibly low and was almost completely offset by the state, have led to pollution with pesticides and mineral fertilizers not only in the rivers of the Black Sea region of Ukraine, but also in the water of the Black and Azov Seas. Application of significant amounts of herbicides, pesticides, and fertilizers was required not only for paddy culture; they were in use also for grain and vegetable rotation, as well as for grape groves. Therefore, one has no reasons to charge the paddy culture industry with all the existing ecological problems. For example, the state of the Azov Sea is largely dependent on the environmental safety of technologies used for growing of crops, including rice, in a neighboring country, Russia; hence, the water from its southern agricultural areas is discharged into the Azov Sea. Only decades after that, the ecologists started their researches and found out the enormous influence of mentioned substances on fish culture, wild marsh plants, and biodiversity overall [29–32].

Rice is one of the few crops, the intensive production of which requires the use of significant amounts of irrigation water, doses of mineral fertilizers and chemical agents for the protection of crops from pests, diseases, and weeds. Application of significant amounts of pesticides and herbicides, the unreasonable irrigation schedule conditioned the pollution of receiving basins with chemical crop-protecting agents and water conversion everywhere [33–35]. There were the reasons of rice cultivating areas and gross output negative dynamics (Fig. 2, as per Governmental statistics).

In recent days, one of the reasons of reducing rice areas in Ukraine is the problem of weed control in rice crop rotation, especially ruderal rice with colored grain, which appeared more widely in Eastern and Southern Europe [36]. As per systematic attitude, it is the same species that cultivated rice *Oryza sativa L.* All of its forms have negative characteristics such as lodging, fall of grain, poor grain quality, susceptibility to main diseases. On the other hand, list of characteristics allows ruderal rice to be more competitive than cultivated rice. For example, longer seeds dormancy period, stretched germinating and blowing period, rapid growing, vigor of shoot growth. Blotchy ripening and ability to fall of grain immediately after

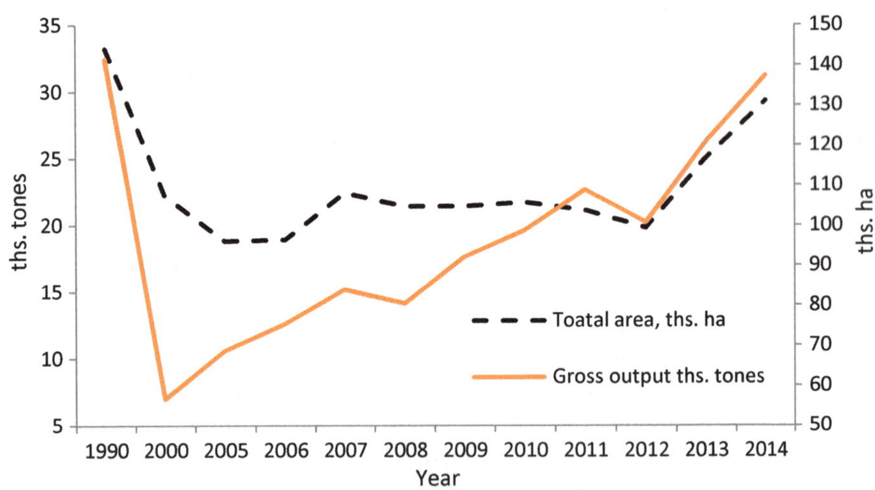

Fig. 2 Dynamics of rice gross output and cultivating area in Ukraine during period 1990–2014

ripening accumulate abnormal stocks of ruderal rice seeds in the field [37–42]. All of mentioned above produce difficulties in rice growing, even though totally infested fields usually removed from rice crop rotation for at least 2–3 years.

This situation is not the isolated case of the late 80s of the XX century; it also emerges in the 60s of the same century. It is noted by experts in connection with unjustified ecologically dangerous technologies, which at that time are already used in irrigation farming and, especially, in the flooded paddy fields. The huge volumes of the freshwater with dissolved chemical agents and fertilizers are discharged into the bays and basins of the Black and Azov Seas. In the 70s of the last century, coastal waters of the southern seas were unacceptably desalted so that the water salt content in the Dzharylhats'ka Gulf in summer reached only 1–2 g/L compared to the natural level of 16–18 g/L [43].

4 Legislation Regulatory

In the explanatory note (1967) to the project of the overall irrigation in the south of Ukraine submitted by the research institute "Ukrdniprovodhosp," scientists and designing engineers forecasted the next scenario: "in view of the continuous rice cultivation in the coastal zone radical measures will be required in order to protect large settlements from flooding which will include liquidation of the resort zone along the Black Sea coast and protection of 32 settlements in the rice cultivation zone …." This forecast has come true to some extent, although such a critical situation could be avoided had these scientists and designing engineers scheduled in their projects some measures for protection of soil fertility, water resources, and environment. Though they have limited themselves to mere verbal warnings, not providing any pro-active approach. And nowadays we need to eliminate these mistakes, deal with ecological challenges, maintain the paddy culture, and provide protection and sustainability of recreational resources of the south region. In connection with the complicated ecological situation in the early 80s in the south region of Ukraine, the Council of Ministers of the USSR, by Resolution No. 166 from 09.04.1976 "On measures to prevent pollution of the basins of the Black and Azov Seas" (paragraph 8) to the Ministry of Agriculture and the Ministry of Water Resources, the Southern Department of VASGNIL was com missioned to conduct the necessary studies and to develop, by 1980, measures aimed at improvement of the environmental status of the south region of Ukraine and abandoning the discharge of water containing pesticides and their compounds from rice irrigation systems into the gulfs and limans of the Black and Azov Seas. Within the scope of these executive directives between 1976 and 1986, Ministry of Water Resources of Ukraine has accomplished reconstruction of 40% of rice irrigation systems. The reconstruction was aimed at the increase of the technical level of the RISs though it did not address the main issues—water and land resources conservation and environment protection. And in the early 90s with the break-up of the Soviet Union, the reconstruction was completely stopped [44, 45].

According to the Water Codex of Ukraine (1995), Chap. 18, "Use of water resources lands. Water protection zones and sanitary protection zones" along the seas and around the sea bays and gulfs "… it is allocated a protected shoreline belt with width of not less than two kilometers from the edge of water." Protected shoreline belts are nature reserves of limited economic activity. Besides, "protected shoreline belt along the seas and around the sea bays and gulfs is a part of the sea sanitary protection zone and can be used only for the construction of sanatoria and other recreational compounds provided with centralized water supply and canalization." According to Art. 93 (Chap. 18) of the aforementioned Water Codex of Ukraine "… in order to protect water objects in the areas of water abstraction for therapeutic and health-promoting needs, it is necessary to establish sanitary protection zones with subdivision into zones of special regime."

Being under the regulation of such Laws of Ukraine as "On the National Program for the Formation of the National Ecological Network of Ukraine for 2000–2020," "On the Nature Reserve Fund of Ukraine" (2000), "On the environment protection" (1996), and the "Water Code of Ukraine" (1995), the Black Sea region territories, including paddy rice culture areas of Ukraine, are provided with four levels of protection from the negative impact of nature management, nevertheless today we recognize these territories as being in the state of ecological depression of varying degrees and also consider them as zone of potential ecological and economic risks. One can confidently assert that the existing legislative and regulatory framework is almost never implemented or fulfilled. This is the reason for these territories being in the state of ecological depression.

The Resolution of the Cabinet of Ministers of Ukraine No. 703 dd. May 18, 1998 "On the sanitary protection of the Skadovsk Medical Zone within the Skadovsky District" strictly prohibits the use of crops growing technologies (e.g., for rice, vegetables, grapes) which involve the use of chemical crop-protecting agents. On the other hand, it is found that this district is more polluted with herbicides, including glyphosate and piretroids compare with neighbor districts [46].

In accordance with the existing legislative and regulatory framework, the width of the sanitary protection zone is 2 km from the shoreline of the Black and Azov Seas, and using this zone imposes stringent requirements on the rice growers and other agricultural producers concerning application of crop cultivation technologies.

Socioeconomic development of the Black Sea region of Ukraine and the special aspects of environmental safety policy formation in the region should take into account the peculiarities of the natural ecological systems and set out effective environment mitigation actions.

The proper quality of the environment should be considered not only as the prerequisite program but also as the result of the region development strategy implementation. Based on this methodological approach, there arise two fundamental requirements: firstly, it is necessary to provide for simultaneous handling of all the socioeconomic, environmental, and economic issues, as well as problems related to the usage of natural resources potential, economic development, and environmental quality management. Secondary, it is necessary to carry out a comprehensive assessment of the programs and projects of the region development, considering the possible ecological and social consequences, including those conferring adverse effect on ecological and economic systems of different development level, from the local to the macroeconomic.

The transition of the region to the environmentally friendly level of development involves the proper balancing of issues connected to the social development, raising the well-being of the people and the economic development of the region on the one hand and natural resources potential on the other hand. When designing policies of the environmentally balanced development of the region, it is necessary to rely on the acceptable demographical, ecological, production, and technical and process design indices.

The analysis of the usage pattern of the natural resources being the most important on the territory of the region (land, water, recreational, and others) has shown that the natural resources potential of the region is the important prerequisite for the socioeconomic development, contributing to the implementation of the economic branches transformation in the region [22].

The study of nature management patterns in the region shows that almost every branch in the region is characterized by an insufficient application of environmentally friendly and resource-saving technologies, the low level of mechanization and automation of production processes, the prevalence of technologies oriented on the intensive use of natural resources, and running of production processes with low coefficient of resource productivity. This situation results in the unjustified large losses of natural resources. The main causes of this situation include high energy intensity of the economy and a significant level of environmental pollution, the impact of extensive agricultural production on the state of ecological and economic systems, the shortage of water resources in a significant part of the country, as well as insufficient level of development of environmental institutions [47].

Adopted in 2003, the Law of Ukraine "On the Fundamentals of National Security of Ukraine" broadly interprets the concept of national security and draws attention to such environmental problems as industrial pollution and the generation of waste, deterioration of water quality, inappropriate use of natural resources, adverse effects of past military activities, as well as environmental hazards of existing obsolete technologies.

The list of external security hazards includes global environmental changes and the import of potentially harmful technologies, and specifically genetically modified organisms. The law also addresses cross-border environmental problems, including issues concerning co-management of cross-border basins and problems related to pollution of the Black and Azov Seas.

5 Short-Term Outlook

To our opinion, the formation of environmentally friendly economic policy for the Black Sea region of Ukraine should involve such main directions as:

- indication of the natural resources potential features and advantages of the region, which distinguish it from the other regions and determine possible directions for development;
- analysis of the natural resources potential status and identification of disadvantages inherent to the region development with a view to their subsequent elimination;
- designing of a strategic plan for the use, restoration, and protection of natural resources on the basis of the top-priority goals of the region's development;
- synchronous reorganization of the institutional management of the region's economy, which involves the creation of institutions oriented toward a market economy, improvement of the regulatory framework concerning use, conservation and reproduction of natural resources. The institutional aspects of the nature management envisage the ecological orientation of the bio-socioeconomic system functioning in the region.

Such aspects of scientific and technological progress as environment protection, environment safety, and resource saving should be prioritized in all areas of the region's economy and used as a guideline for designing of investment and innovation policies of the region. These directions should be considered as one of the top-priority goals on the path to the ecologically balanced nature management, which will lead to overcoming of the resource-ecological crisis and will contribute to the environment improvement.

The introduction of environmentally friendly policies in the region will create conditions for optimal and balanced functioning of the ecology-economic system, which will allow:

- to balance the economic structure of the region;
- to ensure nature management with respect to ecological principles and preservation of the natural environment;
- to form a modern market of competitive agricultural products together with recreational and tourist services;
- to ensure a high social level and increase of well-being of the population in the region;
- to reduce the level of depression of the region and create conditions for non-conflict multidisciplinary use of natural resources potential of the region.

Considering the Black Sea region depression status and the ongoing conflicts of interests arising from multisectoral usage of its main natural resources, as well as the fact that the territory of the region belongs to the zone of potential ecological and economic risk, it is quite logical to conclude that at this stage of socioeconomic development of the region and the implementation of market transformations, there is an urgent need for a fundamentally new sustainable macroeconomic and microeconomic policy. This policy should determine the real courses, methods, means, and effective mechanisms for the transition of the eco-economic or bio-socioeconomic system "economy-society-ecology" to the model of environmentally sustainable development both at the state and regional and local levels.

The implementation of the principles of environmentally sustainable development is particularly important at the regional and local levels, since this is where the state authorities, the public and society understand the social, economic, and resource-ecological problems, see concrete ways and possibilities for their purposeful solution in the best possible way. Designing the principles of environmentally sustainable development requires separate consideration of the characteristics of each region and each separate territory. Currently, each region of Ukraine has its specific territorial production and socio-ecological systems, which are characterized by a certain composition and structure as well as quite large differences.

The model for the environmentally sustainable nature management resides on the concept that economic models of development are closely interrelated with macroeconomic policies and is characterized by criteria defined below:

- the level of social relations development concerning development of their environmental aspect;
- limited natural resources;
- the level of scientific and technological development and the possibility of technology transfer.

In addition, there is an urgent need for detailed assessment the agrarian sector of the economy as for its trends and impact it impose on the ecological state of the recreational resources of the region and the recreation area itself. The results of such an assessment could be further used to increase the effectiveness of activities aimed primarily at resolving contradictions and achieving a balance of interests between different sectors of the economy, developing in the region.

In the context of this study, it is important to address the issue concerning the development of the environmentally sustainable nature management concept for the zones, which are zones polyfunctional purpose and at the same time zones of potential ecological and economic risks (for example, the zone of paddy culture in Ukraine). The dominant idea of the concept is the optimization of nature management within the existing ecological potential and maintaining the balance of the eco-economic system of the region. The concept of environmentally sustainable nature management, developed by the authors of this research, in the context of government-wide and national interests can serve as a powerful tool for effective management decisions on the implementation of environmental policy in general, the identification of priority vectors for the development of territories, taking into account models of their multifunctional use and the balance of interests regarding the optimality of environmental decisions.

From the foregoing, it follows that further economic growth in the regional context and the exit from the ecological and economic crisis is possible only under the following conditions:

- elimination of existing developmental disproportions of the economic-industrial complex of the region;
- development of measures and mechanisms for attracting funds from local authorities to combat the flooding and solving social problems, arising as a result of flooding;
- rational usage of unique recreational potential in the context of environmentally sustainable nature management model of the region;
- design of an effective regional innovation policy, based on the latest scientific and technical achievements;
- development of the model for the investment and innovative development of the region, which will provide conditions for the transformation of scientific research results into resource-saving, science-intensive, low-waste and non-waste technologies, and competitive goods and services;
- dedicated improvement of the financial and credit mechanism of investment and innovative development via transition to program-targeted development management methods that involve the definition of specific tasks, specific research objects, and due dates of research work with the obligatory consideration of global development processes.

In the existing competitive business environment development of an effective ecological-economic nature management demand transition to the business mechanism of state and regional management. This can be achieved through the formation of an effective legislative environment, the use of effective economic instruments, the formation of investment and innovation policies, as well as the executive mechanisms for ensuring the preservation of the environment and socio-economic development of the region.

The negative impact of paddy culture on the environment is predominantly attributable to the discharges of drainage and wastewater outlet containing residues of pesticides, herbicides, and mineral fertilizers beyond the limits of irrigation systems. Nowadays, there is a tendency for decrease of rice yield and production profitability, and the ecological status of the paddy culture and the adjoining off-shore zones has considerably deteriorated. The time is ripe for radical ecological, agricultural, and economic measures.

In the mid-80s of the twentieth century, in Ukraine was launched the process of rice irrigation systems reconstruction. There were made certain steps toward solving the problems concerning the ecological status of both rice irrigation systems and the Black Sea and Azov Seas basins in general. However, the reconstruction, which took the form of partial improvement of rice irrigation systems (channels lining, improvement of hydro-technical utilities, arrangement of closed collector-drainage network, improvement of rice cultivation technology, use of drainage and wastewater for return flow irrigation), did not radically reduce the sharpness of environmental problems in the southern region of Ukraine.

In recent years, the Ukrainian water industry is setting its sights on transition to ecologically sound agricultural production, obtaining stable yields of environmentally safe agricultural products, reduction of electricity costs, saving of water and land resources and preservation of soil fertility. This course is also accompanied with improvement of quality and reliability of scientific research results, as well as the introduction of new rice production technologies and new generations of rice irrigation systems.

New technically upgraded rice irrigation systems should meet the following requirements:

- be environmentally friendly;
- to ensure electricity conservation, as well as water and land resources preservation and reproduction;
- to maintain the hydrogeological and reclamation state of irrigated land in accordance with paddy culture requirements;
- to ensure the preservation of soil fertility;
- to prevent pollution of agricultural and adjacent territories and marine areas;
- to promote the introduction of environmentally friendly technologies with reduced application of herbicides for the cultivation of rice and related crops.

6 Conclusions

The model of rational nature management in the paddy culture area of Ukraine should be based on the concept of sustainable nature management and provide for integrated development of the rice industry and recreational complex with simultaneous preservation of natural resources potential and environment of the region:

- provide the perspective plan for development of the paddy culture branch on the basis of reconstruction of existing rice irrigation systems into closed-type paddy field return flow irrigation systems and the introduction of new environmentally friendly technologies for the cultivation of rice and related crops;
- provide the road map for the rational use of recreational natural resources of the Black and Azov Seas, establishment of a recreational complex and appropriate infrastructure, taking into account the requirements of sustainable nature management.

Realization of these requirements is possible only with consideration of all the interrelations and factors of mutual influence inherent to components of natural resources potential of the paddy culture area, which can be used now and in the future for improvement of the population living conditions, ecological well-being, and integrated development of the region.

References

1. Bambaradeniya CNB, Edirisinghe JP, De Silva DN, Gunatilleke CVS, Ranawana KB, Wijekoon S (2004) Biodiversity associated with an irrigated rice agro-ecosystem in Sri Lanka. Biodivers Conserv 13:1715–1753. https://doi.org/10.1023/B:BIOC.0000029331.92656.de
2. Ilieva V, Markova-Ruzdik N, Andreevska D, Andov D (2010) Breeding and evaluation for improved rice varieties in Macedonia. Plant Sci 47(1):17–22
3. Martínez-Eixarch M, Curcó A, Ibáñez C (2017) Effects of agri-environmental and organic rice farming on yield and macrophyte community in Mediterranean paddy fields. Paddy Water Environ 15(3):457–468. https://doi.org/10.1007/s10333-016-0563-x
4. Petrenko V, Liubich V, Bondar V (2017) Baking quality of wheat grain as influenced by agriculture systems, weather and storing conditions. Rom Agric Res 34:69–76
5. Zhang B (1997) Total factor productivity of grain production in the former Soviet Union. J Comp Econ 24:202–209. https://doi.org/10.1006/jcec.1997.1417
6. Granovska LM, Polukhov AY (2003) Influence of land reform on the efficiency of rice production in the South of Ukraine. Vodni Resursy ta Melioratsiya 27:60–65
7. Bambaradeniya CNB (2000) Ecology and biodiversity in an irrigated rice field ecosystem. Doctoral dissertation, Sri Lanka
8. Concepción E, Díaz M, Baquero R (2008) Effects of landscape complexity on the ecological effectiveness of agri-environment schemes. Landscape Ecol 23:135–148. https://doi.org/10.1007/s10980-007-9150-2
9. Belousov IE, Paraschenko VN (2013) Implementation of effective fertility of meadow chernozem soil under rice with adaptive use of mineral fertilizers. Risovodstvo 22:59–65
10. Budelsky RA, Galatowitsch SM (2000) Effects of water regime and competition on the establishment of a native sedge in restored wetlands. J Appl Ecol 37(6):971–985. https://doi.org/10.1046/j.1365-2664.2000.00540.x
11. Tabot PT, Adams JB (2013) Ecophysiology of salt marsh plants and predicted responses to climate change in South Africa. Ocean Coast Manag 80:89–99. https://doi.org/10.1016/j.ocecoaman.2013.04.003
12. Takanose Y, Ishida S, Kudo N, Kamitani T (2013) Effects of tillage and irrigation on the occurrence and establishment of native wetland plant species in fallow paddy fields. Paddy Water Environ 11(1–4):45–58. https://doi.org/10.1007/s10333-011-0292-0
13. Janaguiraman DM, Ramadass R, Devi D (2003) Effect of salt stress on germination and seedling growth in rice genotypes. Madras Agric J 90(1):50–53
14. Zeng L, Shannon MC (2000) Salinity effects on seedling growth and yield components of rice. Crop Sci 40(4):996–1003. https://doi.org/10.2135/cropsci2000.404996x
15. Morozov RV (2010) Trends in rice cultivation in the south of Ukraine. Bull Kamyanets-Podilsky Natl Univ 12(3):152–156
16. Dudchenko VV, Kropivko MF, Morozov RV, Chekamova OI (2009) Zoning of rice cultivating areas. Vodni Resursy ta Melioratsiya 33:81–95
17. Mendus SP (2010) Improvement of drainage of rice systems as a means of reducing the risk of adverse effects of artificial objects on the surrounding shedding area (on the example of the Kilia rice irrigation system, Odessa region). Zemlerobstvo 12(1):17–19
18. Stashuk VA, Rokochinsky AM, Granovska LM, Vozhegova RA (2013) History of construction and efficiency of usage of rice irrigation systems in Ukraine. Vodni Resursy ta Melioratsiya 35:9–13
19. Wuepper D, Heissenhuber A, Sauer J (2017) Investigating rice farmers' preferences for an agri-environmental scheme: Is an eco-label a substitute for payments? Land Use Policy 64:374–382. https://doi.org/10.1016/j.landusepol.2017.03.014
20. Dudchenko VV (2006) Environmental safety and the rice cultivation. Naukoviy visnyk NUWR 11(1):4–8

21. Iizumi T, Furuya J, Shen Z, Kim W, Okada M, Fujimori S, Nishimori M (2017) Responses of crop yield growth to global temperature and socioeconomic changes. Sci Rep 7(1):7800. https://doi.org/10.1038/s41598-017-08214-4
22. Koch M (2015) Climate change, capitalism and degrowth trajectories to a global steady-state economy. Int Crit Thought 5(4):439–452. https://doi.org/10.1080/21598282.2015.1102078
23. Casanova D, Goudriaan J, Català MM, Withagen JCM (2002) Rice yield prediction from yield components and limiting factors. Eur J Agron 17:41–61. https://doi.org/10.1016/S1161-0301(01)00137-X
24. Goncharova JK, Kharitonov EM (2009) Indicators of productivity of domestic rice varieties at high temperatures due to global climate change. Sel'skokhozyaistvennaya Biol 1:16–20
25. Vozhegova R (2004) Varieties and directions of rice breeding in Ukraine. In: Challenges and opportunities for sustainable rice-based production systems. Università degli Studi di Torino, Italy, pp 13–15
26. Levins R (1990) The struggle for ecological agriculture in Cuba. Capitalism Nat Social 1 (5):121–141. https://doi.org/10.1080/10455759009358419
27. Kovalev SV, Kozihkurt ME, Kozichkurt SM (2004) Necessity and possibility of preservation of rice ecosystems of Ukraine. Bull NUVGP 28:41–49
28. Mauerhofer V, Ichinose T, Blackwell BD, Willig MR, Flint CG, Krause MS, Penker M (2018) Underuse of social-ecological systems: a research agenda for addressing challenges to biocultural diversity. Land Use Policy 72:57–64. https://doi.org/10.1016/j.landusepol.2017.12.003
29. Abdullah AR, Bajet CM, Matin MA, Nhan DD, Sulaiman AH (1997) Ecotoxicology of pesticides in the tropical paddy field ecosystem. Environ Toxicol Chem 16:59–70. https://doi.org/10.1002/etc.5620160106
30. Fernando CH (1993) Rice field ecology and fish culture—an overview. Hydrobiologia 259:91–113
31. Katayama N, Baba YG, Kusumoto Y, Tanaka K (2015) A review of post-war changes in rice farming and biodiversity in Japan. Agric Syst 132:73–84. https://doi.org/10.1016/j.agsy.2014.09.001
32. Maltchik L, Stenert C, Batzer DP (2017) Can rice field management practices contribute to the conservation of species from natural wetlands? Lessons from Brazil. Basic Appl Ecol 18:50–56. https://doi.org/10.1016/j.baae.2016.10.002
33. Roger PA, Simpson I, Oficial R, Ardales S, Jimenez R (1994) Effects of pesticides on soil and water microflora and mesofauna in wetland rice fields: a summary of current knowledge and extrapolation to temperate environments. Aust J Exp Agric 34(7):1057–1068. https://doi.org/10.1071/EA9941057
34. Ueji M, Inao K (2001) Rice paddy field herbicides and their effects on the environment and ecosystems. Weed Biol Manage 1(1):71–79. https://doi.org/10.1046/j.1445-6664.2001.00002.x
35. Willocquet L, Elazegui FA, Castilla N, Fernandez L, Fischer KS, Peng S, Savary S (2004) Research priorities for rice pest management in tropical Asia: a simulation analysis of yield losses and management efficiencies. Phytopathology 94(7):672–682. https://doi.org/10.1094/PHYTO.2004.94.7.672
36. Eleftherohorinos IG, Dhima KV, Vasilakoglou IB (2002) Interference of red rice in rice grown in Greece. Weed Sci 50(2):167–172. https://doi.org/10.1614/0043-1745(2002)050%5b0167:IORRIR%5d2.0.CO;2
37. Delatorre CA (1999) Dormência em sementes de arroz vermelho. Cienc Rural 29(3):565–571
38. Estorninos LE, Gealy DR, Gbur EE, Talbert RE, McClelland MR (2005) Rice and red rice interference. II. Rice response to population densities of three red rice (*Oryza sativa*) ecotypes. Weed Sci 53(5):683–689. https://doi.org/10.1614/ws-04-040r1.1
39. Ferrero A (2010) Weedy rice, biological features and control. Weed Management for Developing Countries. FAO Corporate Repository. Available on http://www.fao.org/docrep/006/Y5031E/y5031e09.htm

40. Fogliatto S, Vidotto F, Ferrero A (2012) Morphological characterization of Italian weedy rice (*Oryza sativa*) population. Weed Res 52(1):60–69. https://doi.org/10.1111/j.1365-3180.2011.00890.x
41. Shivrain VK, Burgos NR, Scott RC, Gbur EE, Estorninos LE, McClelland MR (2010) Diversity of weedy red rice (*Oryza sativa L.*) in Arkansas, USA in relation to weed management. Crop Prot 29(7):721–730. https://doi.org/10.1016/j.cropro.2010.02.010
42. Singh S, Singh G, Singh VP, Singh AP (2005) Effect of establishment methods and weed management practices on weeds and rice in rice-wheat cropping system. Indian J Weed Sci 37 (1):51–57
43. Yurchenko AI (2016) Measures to increase the degree of environmental safety of rice irrigation systems. Ekologia 38:133–142
44. Vantsovsky AA, Vozhegov SG, Vozhegova RA (2004) Technology of rice cultivation due to requirements of environmental protection in Ukrainian farms. Risovodstvo 13:60–77
45. Zorodychev NV, Levinska BV, Dedova EB, Ochirova EN (2012) Ecological and energetic effectiveness of rice agro-landscapes in Sarpinska area. Plodorodiye 2:31–38
46. Zotikov VI, Bobkov SV (2010) Characteristic of environmental pollution in Kherson region. Ekologia 32:17–19
47. Tsuchiya K, Hagihara K (2017) Agricultural landscape: farmland abandonment and direct payments. In: Labor forces and landscape management. Springer, Singapore, pp 83–98. https://doi.org/10.1007/978-981-10-2278-4_5

Development of a New Formulation for Onychomycosis Treatment Using Furvina® as an Active Pharmaceutical Ingredient

Zenia Perez-Rodriguez, Yaset Rodríguez-Rodríguez, Zenaida Rodriguez-Negrin, Reinaldo Molina-Ruiz, Ricardo Medina-Marrero and Evys Ancede-Gallardo

Abstract The currently available treatments for onychomycosis are limited due to the absence of a formulation allowing the effective penetration of the active pharmaceutical ingredient (API). In this research, a chemical–pharmaceutical study over a new formulation for the treatment of onychomycosis, based on 2-bromo-5-(2-bromo-2-nitrovinyl) furan (Furvina) composed by 0.25% of this API, 1% urea as keratinizing agent and 98.75% benzyl alcohol as pharmaceutical vehicle, was performed. This tincture proved to be active against *Candida albicans*. In addition, an UV/Vis spectrophotometric analytic technique with a maximum absorption at 372 nm was developed and validated. This method was specific, precise, accurate, and linear in the range of 2–28 mg/L. The detection and quantification limits for this technique were 0.6567 and 0.6636 mg/L, respectively.

Keywords Onychomycosis · Furvina · Spectrophotometric method UV/Vis · Validation

Highlights

- A new formulation for onychomycosis treatment having Furvina as an API with broad-spectrum antifungal activity has been developed.
- A new spectrophotometric method was developed and validated for quantifying Furvina in the tincture.

Z. Perez-Rodriguez (✉) · Y. Rodríguez-Rodríguez · Z. Rodriguez-Negrin
R. Molina-Ruiz · R. Medina-Marrero · E. Ancede-Gallardo
Centro de Bioactivos Químicos (CBQ), Universidad Central "Marta Abreu" de Las Villas, Carretera a Camajuaní Km 5½, Santa Clara CP 54830, Villa Clara, Cuba
e-mail: zperez@uclv.edu.cu

R. Cárdenas et al. (eds.), *Proceedings of the 2nd International Conference on BioGeoSciences*, https://doi.org/10.1007/978-3-030-04233-2_17

1 Introduction

Pharmaceutical companies must insert new formulations into the market in order to ensure their growth. However, the costs of drug research and development (R&D) are very high and promising results are not always obtained [1]. The active pharmaceutical ingredient (API) and its formulation have to undergo several phases of testing, prior to its insertion as a new drug. Consequently, the authorization for its commercialization may take over 10–15 years. Besides, many of these products do not arrive at their final evaluation stage and have to be abandoned [2].

Tinea unguium, onychomycosis or ringworm of the nails, is a disease caused by fungi [3]. Yeasts of the genus *Candida* produce damages on fingernails and nail folds without predominance to any of the fingers. This infection (onychomycosis) is unpleasant to patients due to the incidence of this disease in social relationships and especially in the work related to food handling [4]. Moreover, onychomycosis has a worldwide distribution and high incidence, affecting approximately 10% of the general population and up to 50% of people aged over 60 years [5, 6].

Many formulations to treat onychomycosis are recognized in the Central American market [2], but they have high concentrations of APIs and require lengthy treatment periods. The process of new drug development passes through several well-defined phases [7–9]. The first steps are the formulation studies and the chemical–pharmaceutical characterization of new drugs. Furthermore, chemical methods for the determination and quantification of APIs in the formulations are needed in the characterization process.

Furvina® or G-1 is a chemical compound named 2-bromo-5-(2-bromo-2-nitrovinyl)-furan. This substance has a broad antimicrobial spectrum [10, 11], and it is used in different formulations as an antifungal and antibacterial agent. It is used in the following formulations: Dermofural®, Queratofural®, and Furvinol®, which are used in human and veterinary health; and Vitrofural®, which is utilized as chemical sterilant in vitroplant production [12]. The aim of this research is to develop a new formulation for onychomycosis treatment using Furvina® as an API.

2 Materials and Methods

The Centro de Bioactivos Químicos (CBQ) it has been proposed a new drug formulation for the treatment of onychomycosis. This formulation is based in a tincture with Furvina® (commercial name) as API, urea, and benzyl alcohol as keratinizing agent and adjuvant, respectively.

2.1 Materials

Pure samples, Furvina® standard (99.73% produced in CBQ), ethanol, urea, benzyl alcohol, and other reagents of analytical grade, were purchased from Merck Millipore, Germany.

2.2 Preparation of Tincture

In the process of selecting a suitable solvent for making a tincture formulation, several aspects were taken into account. It was chosen a solvent that has reported that use as a pharmaceutical excipient [13] and must meet the requirement of being chemically inert to the other ingredients of the formulation. As a last aspect, the excipient chosen must have a moderate polarity that is capable of dissolving both the active ingredient and urea. In this selection process, benzyl alcohol was elected, with a previous experimental verification that has the chemical and physical requirements to be used in our tincture formulation.

The tincture solution was prepared using 0.25% 2-bromo-5-(2-bromo-2-nitrovinyl) furan (Furvina®) as an API, 1% urea as a keratinizing agent, and 98.75% benzyl alcohol as a pharmaceutical vehicle. The mixture was placed in an ultrasonic bath for 15 min to achieve its complete dissolution. Then, the same solvent was subsequently added to this solution in a 100 mL volumetric flask to obtain a final concentration of Furvina of 0.25 μg/mL.

2.3 Equipment

The main instruments used were a UV/Vis spectrophotometer with one cm matched quartz cells (model Thermo), an electronic balance (Sartorius CP 225) and an ultrasonic cleaner (BRANSON 5510). The glassware used in each procedure was soaked overnight in a mixture of chromic acid and sulfuric acid, rinsed thoroughly with double distilled water, and dried with hot air in an oven before use. The absorption spectra of reference and test solution were carried out in quartz cells (one cm) over the range of 190–500 nm.

2.4 Antifungal Activity of the Tincture

The antifungal activity of the tincture was assessed by the agar diffusion method. Wells were puncture using a sterile cork borer from Sabouraud dextrose agar, previously seeded with *Candida albicans* ATCC 10231 as a test organism. Wells were filled with 15 μL of the tincture, whereas wells containing all formulation ingredients but API served as controls. The diameter of inhibition was determined after 48 h of incubation at 30 °C.

2.5 Method Development

Selection of the analytical wavelength to determine Furvina in the tincture
The tincture was scanned in the range of 200–400 nm wavelength in order to obtain the maximum absorbance.

Selection of the analytical concentration range and preparation of calibration curves
Different aliquots from the tincture stock solution were pipetted into 25-mL volumetric flasks to obtain different standard solutions of the tincture. The absorbance of each concentration was measured three times in different days to obtain the calibration curve at 372 nm.

2.6 Method Validation

The analytical method for quantification of Furvina in the tincture was validated according to the guidelines of The British Pharmacopoeia [14], the International Conference on Harmonization (ICH) [15], and Centro para el Control Estatal de la Calidad de los Medicamentos (CECMED) (in Spanish) [16]. The validation parameters selected were specificity, linearity, veracity, intermedia precision, limit of detection (LOD), and limit of quantification (LOQ).

The tincture was dissolved in ethanol due to its solubility in this solvent, which was used during the validation process.

Specificity
The evaluation of specificity of this method was carried out according to the following procedure: Tincture ingredients except API (i.e., urea and benzyl alcohol) were dissolved in ethanol, each in a different flask, taking this as a reagent blank. A tincture solution was prepared in parallel. Aliquots from above solutions were placed in a spectrophotometer, and the corresponding spectra were registered in the range of 190–500 nm to determine whether the method was specific for quantifying Furvina in the tincture [16].

Linearity
Standard solutions of Furvina were prepared at eight concentrations (2, 4, 8, 12, 16, 20, 24, and 28 mg/L), and the corresponding absorbance values were determined in triplicate at 372 nm in three different days. The obtained results were used to evaluate the linearity of the method by means of different recognized tests [17, 18].

Limit of detection (LOD) and limit of quantification (LOQ)
The limit of detection (LOD) and limit of quantification (LOQ) were determined according to the method of extrapolation to zero, described by the Spanish Association of Pharmaceutical Industry [19]. See Eqs. (1) and (2).

$$\text{LOD} = \frac{\bar{y}_{bl} * 3S_{bl}}{b\sqrt{n}} \tag{1}$$

$$\text{LOQ} = \frac{\bar{y}_{bl} * 10S_{bl}}{b\sqrt{n}} \tag{2}$$

where

S_{bl}: the standard deviation of the analytical response,
b: the slope of the dynamic linear range,
$\bar{y}_{bl}$: the intercept of the straight line between $\text{Abs}_{\text{average}}$ versus c (mg/L),
n: number of concentrations studied.

Precision in terms of repeatability and intermediate precision
Repeatability

Three tincture solutions with different levels of Furvina concentrations were prepared to evaluate the repeatability of the method. Six replicates of each concentration were measured under homogeneous conditions of analysis. After that, the mean concentration, standard deviation, and the coefficient of variation of repeatability (CV_r) were calculated.

Intermediate precision

The intermediate precision (within-laboratory variation) was determined by three analysts at the same day, using the same concentration levels as previously described. Then, the standard deviation ($S_{\text{between/day}}$), the coefficient of variation between days ($\text{CV}_{\text{between/day}}$), theoretical variation coefficient according to Horwitz (Horwitz CV%), and the statistician Cochran (C) were calculated for the evaluation of this test.

Veracity

The veracity of the method for the recovery test was determined according to the following procedure. Nine solutions or references with known concentrations were prepared at 20 mg/L (this concentration is located at an intermediate point of the calibration curve). The average value found was compared with the expected value for each case, through the calculation of recovery [20]. See Eq. (3).

$$\frac{\bar{C}_X}{C_{\text{THEOR}}} = \frac{R}{100\%} \tag{3}$$

where

$\bar{C}_X$: experimental mean concentration,
C_{THEOR}: theoretical concentration or expected value,

R: regained value obtained.

A hypothesis test was performed using Student's t-distribution, raising the null hypothesis:

H_0: $R_m = 100$, and as an alternative hypothesis:
H_1: $R_m \neq 100$, to show whether or recovered medium obtained was not significantly different from 100; for this expression is used: See Eq. (4)

$$t_{exp} = \frac{|100 - R_m|\sqrt{n}}{CV} \quad (4)$$

where

R_m % average recovery,
n number of determinations,
CV coefficient of variation % recovery.

The experimental value of t is compared with the value of the table t, $t_{[(1-\alpha),\ (n-1)]}$, where $\alpha = 0.05$.

In assessing the truthfulness, the following criteria were considered:

- If the average recovery value was in the range of $97\% < R_m < 103\%$, it was considered statistically equal to C_{THEOR}, and consequently, the method was accurate.
- If the value of $t_{exp} < t_{tab}$, there were no significant differences between R_m 100, and a good veracity of the method was established.

3 Results and Discussion

3.1 Microbiologic Evaluation of the Pharmaceutical Preparation

Figure 1 shows the activity of the tincture having Furvina as an API against *Candida albicans* ATCC 10231. It can be appreciated that the tincture (left plates) had stronger antifungal activity than the solvent (right plates) against the test organism, which indicates that the API retained its activity in this formulation. The activity shown by the solvent (right plates) is not surprising as long as the antimicrobial activity of benzyl alcohol is well known [21].

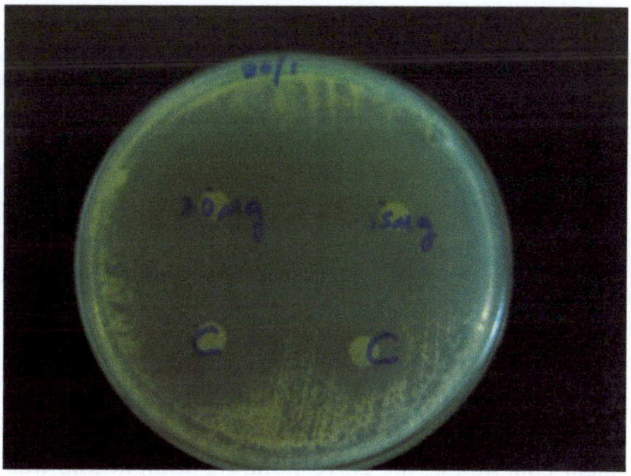

Fig. 1 Evaluation of the tincture against *Candida albicans*. Left and right plates show the activity of the tincture and solvent, respectively

3.2 Selection of the Analytical Wavelength

The solution containing 0.25 μg/mL of Furvina was prepared and scanned over the range of 200–400 nm against ethanol as a blank, using a Thermo UV spectrophotometer (Fig. 2). The maximum wavelength obtained in the graph was considered as the maximum UV absorption for the tincture.

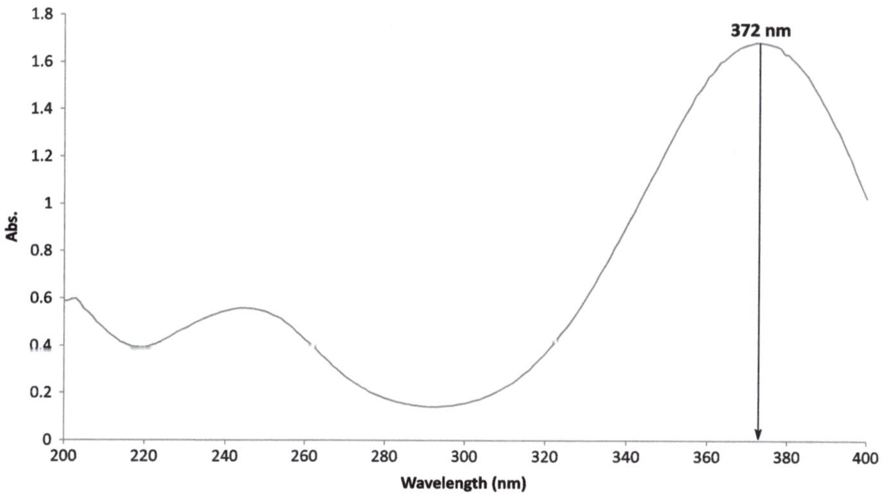

Fig. 2 UV/Vis spectrum of Furvina tincture in ethanol (λ_{max} = 372 nm)

3.3 Validation

Specificity

Furvina samples, urea, and benzyl alcohol, each dissolved in ethanol in different flasks, were taken as blank reagents. A scan was performed from 190 to 500 nm in order to determine whether the method was specific and able of quantifying Furvina in the tincture [5]. It was shown that tincture components do not interfere with Furvina determination, because their absorbance maxima are at lower wavelengths than that of Furvina (Fig. 3).

Linearity

The results about the linearity test for a Furvina range of 2–28 mg/L versus the corresponding absorbance are shown in Table 1. For this concentration range, the linear equation of the corresponding straight line was $y = 0.0589x - 0.0138$, which had a high value of the correlation coefficient (R^2) equal to 0.996. The method also met all the criteria evaluated, indicating that the analytical technique is linear (see Table 1). In this sense, there is a high correlation between absorbance and concentration of Furvina (observed values of r and R^2). Furthermore, the angular coefficient (n) value of 1.0069 indicates that the slope is cut at an angle of 45°. The quality coefficient is a new chemometric tool used by several researchers [16–18] to evaluate the linearity of calibration curves [22]. The quality coefficient determined is lower than the maximum threshold which confirms that there is a high correlation between the variables' absorption and concentration.

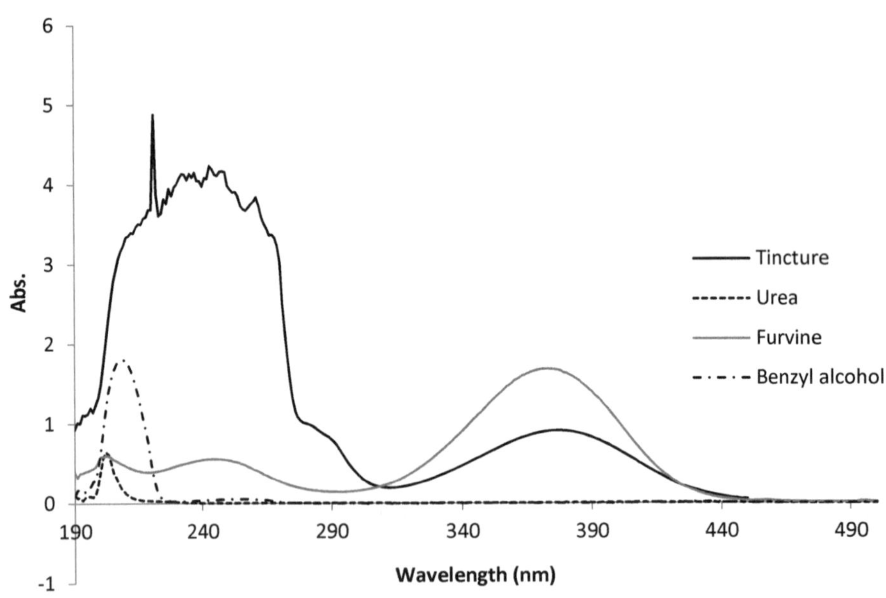

Fig. 3 UV/Vis spectrum of a tincture sample and of each of its components (λ_{max} = 372 nm)

Table 1 Results of the linearity test to determine Furvina at 372 nm, through a spectrophotometric method within the concentration range of 2–28 mg/L

Parameter	Experimental value	Acceptance requirements
Number of data (*n*)	72	–
Linearity range (mg/L)	2–28	–
Correlation coefficient (*r*)	0.999	$r \geq 0.99$
Coefficient of determination (R^2)	0.996	$r^2 \geq 0.98$
Angular coefficient (*n*)	1.0069	0.9–1.1
Slope (*b*)	0.0589	–
Intercept (*a*)	−0.0138	–
Confidence interval of (*a*)	(−0.0287–0.0011)	Containing at 0
Confidence interval of (*b*)	(0.0580–0.0597)	It is different from 0
Relative standard deviation of the slope of the regression line (Sb_{rel}) (%)	0.76	$Sb_{rel} \leq 2$
Coefficient of variation of response factors CV_f (%)	4.06	$CV_f \leq 5$
F-Fisher test	17424.7 > $1.13E^{-85}$	$F > F_{critical}$
Quality coefficient (QC) (%)	4.01	QC ≤ 5

Limit of detection (LOD) and limit of quantification (LOQ)
The limit of detection (LOD) and limit of quantification (LOQ) which represent the sensitivity of the proposed method were determined. The LOD value obtained was 0.0687 μg/mL and the LOQ value obtained was 0.0756 μg/mL, which indicate the high sensitivity of the proposed method.

Precision in terms of repeatability and intermediate precision
The coefficient of variation CV (%) obtained for each concentration level was lower than 3%. This result is acceptable for physical–chemical tests. Therefore, the method is precise in terms of repeatability [16] (Table 2).

The results obtained for the study of intermediate precision of Furvina concentration, based on 18 determinations at low, medium, and high concentration levels by three different analysts using the experimental technique UV/Vis, are shown in Table 3.

Table 2 Experimental values of the repeatability test at low, medium, and high concentration levels

Concentration level	Low *c* (Furvina) (mg/L)	Medium *c* (Furvina) (mg/L)	High *c* (Furvina) (mg/L)
Average	2.0494 ± 0.03	18.2895 ± 0.01	29.7963 ± 0.11
Standard deviation	0.03273	0.00877	0.10835
CV (%)	1.597	0.048	0.364

Table 3 Values of a sample analyzed by different analysts at low, medium, and high concentration levels

Item	Low concentration level			Medium concentration level			High concentration level			Acceptance requirements
	Analyst 1	Analyst 2	Analyst 3	Analyst 1	Analyst 2	Analyst 3	Analyst 1	Analyst 2	Analyst 3	–
Average	2.2232	2.0496	2.0948	18.1338	18.2537	18.2895	28.9906	29.7693	29.7422	–
Standard deviation	0.04648	0.03273	0.05453	0.00931	0.02558	0.00877	0.20551	0.10835	0.28520	–
CV (%)	2.091	1.597	2.603	0.051	0.140	0.048	0.709	0.364	0.959	CV% < 3%
$C_{calculated}$	0.4792 < 0.707			0.4792 < 0.707			0.4792 < 0.707			$C_{calculated} < C_{critic}$
$CV_{between/day}$	2.50 < 9.67			3.18 < 10.40			9.38 < 14.41			$CV_{between/day} < CV_{Horwitz}$

Table 4 Veracity of a reference solution of Furvina of known concentration (20 mg/L)

Parameters	Conc (mg/L)	Acceptance criterion
Experimental average concentration	20.1022	
Known concentration C_{THEOR}	20	
Average value of the % recovered R_m	100.5%	97–103%
CV recovered (%)	0.26	
t_{exp}	−5.83	
$t_{(9-1;0.05)}$	1.8595	

Variations of repeatability coefficients for each analyst were less than 3% in all cases. It was also observed that the coefficient of variation between/day approached the theoretical value of the reproducibility of Horwitz [23], when the concentration levels were increased.

Veracity

The evaluation of the veracity of this method by repetitive analysis of a reference solution is shown in Table 4.

The percentage of recovery was in the range of 97–103%, which is reported as an acceptable criterion for this parameter [23]. The value of the statistic *t*-Student was lower than the *t*-tabulated (t_{tab}) for a 95% confidence level. This guarantees the veracity of the method, which is supported by no significant differences between the average experimental and expected values. These results confirm that the UV/Vis method for determining Furvina® is reliable under the conditions studied.

The results of the validation parameters showed that the proposed method was simple, economic, sensitive, precise, reliable and, consequently, can be used for quantification of Furvina® in the tincture.

4 Conclusions

It developed a new formulation for onychomycosis treatment using Furvina® as an API, which proved active against *Candida albicans*. Also, the UV/Vis spectrophotometric method developed for quantification of Furvina® in the tincture was simple, economic, sensitive, linear, precise, and reliable. The proposed method was based on the light absorption in the UV/Vis region using benzyl alcohol as a solvent, and the maximum absorbance was at 372 nm. Besides, it was linear in the Furvina concentration range of 2–28 μg/mL. Further, the limit of detection (LOD) and limit of quantification (LOQ) were 0.0687 and 0.0756 μg/mL, respectively. The results obtained in terms of intermediate precision and truthfulness showed that the method has good precision and veracity.

Acknowledgements This study was supported by the Centro de Bioactivos Químicos (CBQ) through the Institutional Project 9656. The authors are indebted to MSc. Amalia María

Calvo, MSc. Milagro García Bernal and Marlen Casanova González for their technical assistance and valuable help.

Conflict of Interest Authors declare no conflict of interest regarding this publication.

References

1. Negrin ZR, de Rojas Perez M, Lovillo MP, Torres AM. Caracterización y cuantificación del compuesto resultante de la degradación por la luz del 2-(2-nitrovinil)-furano en estado sólido [Characterization and quantification of the resulting compound of the degradation by the light of the 2 (2-nitrovinil)-fura]
2. DiMasi JA, Hansen RW, Grabowski HG (2003) The price of innovation: new estimates of drug development costs. J. Health Econ 22:151–185. https://doi.org/10.1016/s0167-6296(02)00126-1
3. Welsh O, Vera-Cabrera L, Welsh E (2010) Onychomycosis
4. Ballesté R, Mousqués N, Gezuele E (2003) Onicomicosis: Revisión del tema. Rev Méd Urug 19:93–106
5. Scher RK, Nakamura N, Tavakkol A (2014) Luliconazole: a review of a new antifungal agent for the topical treatment of onychomycosis. Mycoses 57:389–393. https://doi.org/10.1111/myc.12168
6. Conti-Díaz IA (1964) Mycological study of 85 cases of onychopathy. An Fac Med Univ Montevideo. cabdirect.org
7. Carrillo-Muñoz AJ, Tur-Tur C, Hernández-Molina JM, Santos P, Cárdenes D, Giusiano G (2010) Antifúngicos disponibles para el tratamiento de las micosis ungueales. Rev Iberoam Micol 27:49–56
8. de Bedout C, Tabares A, Cano L, Restrepo A (2015) Comportamiento de los agentes etiológicos de las onicomicosis en un laboratorio de micologia de referencia (Medellín 1994–2003)
9. Arenas R (1990) Las onicomicosis. Aspectos clínico-epidemiológicos, micológicos y terapéuticos. Gac Med Méx. bases.bireme.br
10. Fabbretti A, Brandi L, Petrelli D, Pon CL, Castanedo NR, Medina R, Gualerzi CO (2012) The antibiotic Furvina® targets the P-site of 30S ribosomal subunits and inhibits translation initiation displaying start codon bias. Nucleic Acids Res. academic.oup.com
11. Sifontes-Rodríguez S, Monzote-Fidalgo L, Castañedo-Cancio N, Montalvo-Álvarez AM, López-Hernández Y, Diogo NM, Infante-Bourzac JF, Pérez-Martín O, Meneses-Marcel A, García-Trevijano JA, Cabrera-Pérez MÁ. The efficacy of 2-nitrovinylfuran derivatives againstLeishmania in vitro and in vivo. SciELO Brasil
12. Morales-Moreno Y, Medina-Marrero R, Garcia-Bernal M, Casanova-González M, Rodríguez-Pérez R, Fernández-López N, Guerra-Martínez M, Gómez-Marrero R, Rojas-Hernández N (2014) Actividad in vitro de furvina frente a bacterias Gram negativas multirresistentes. Rev CENIC Cienc Biol. redalyc.org
13. Rowe RC, Sheskey PJ, Weller PJ (2006) Handbook of pharmaceutical excipients
14. Beckett AH, Stenlake JB (1988) Practical pharmaceutical chemistry. Athlone Press
15. Branch SK (2005) Guidelines from the international conference on harmonisation (ICH). J Pharm Biomed Anal. Elsevier
16. Centro para el Control Estatal de la Calidad de los Medicamentos (CECMED) (2007) Regulación No. 41-2007: Validación de métodos analíticos. Ámbito Regulador

17. Gomes GC, Salgado HRN (2005) Validation of UV spectrophotometric method for determination of lomefloxacin in pharmaceutical dosage form. Acta Farm Bonaer 24:406
18. Castro Cels M, Gascón Fora S, Pujol Forn M, Sanc Roca JM, Vicente Plas L (1989) Validación de métodos analíticos. Asociación Española de Farmaceúticos de la Industria
19. D'Este P (2002) The distinctive patterns of capabilities accumulation and inter-firm heterogeneity: the case of the Spanish pharmaceutical industry. Ind Corp Change. academic.oup.com
20. Pérez M (2008) Desarrollo y validación de técnicas analíticas por HPLC y SQV, para determinar el contenido de UC-244
21. Castelli MV, Cozzi MV, López SN, Derita MG. Búsqueda de agentes conservantes a partir de productos naturales bioactivos, pp 63–67
22. De Beer J, Naert C, Deconinck E (2012) The quality coefficient as performance assessment parameter of straight line calibration curves in relationship with the number of calibration points. Accredit Qual Assur 17:265–274. https://doi.org/10.1007/s00769-011-0871-1
23. González A, Herrador M (2007) A practical guide to analytical method validation, including measurement uncertainty and accuracy profiles. TrAC Trends Anal Chem. Elsevier

Multipurpose Sensor Network for Electromagnetic Radiation Monitoring

Vladimir Mochalov and Mikhail Bersenev

Abstract The paper discusses the process of a multipurpose sensor network for electromagnetic radiation monitoring (MSNERM) construction. We propose a functional scheme for MSNERM construction, which includes a functional block to synthesize the MSNERM structure, using the existing authorial multi-agent bio-inspired algorithm for synthesizing the sensor network structure. A functional scheme of the MSNERM endpoint functional node is given, and an example of its software and hardware implementation is proposed based on minicomputer, satellite time synchronization GPS/GLONASS module, eight-channel signal recording device in the frequency range up to 100 kHz, software-defined radio module, FPGA, antenna switch, antennas, etc.

Keywords Sensor networks · Electromagnetic radiation · Signal analysis
System analysis

1 Introduction

In this paper, we accept that a sensor network is a distributed network of endpoint functional nodes (F-nodes) which analyze the time-synchronized signals and transmit the collected information (analysis results) to one or several collecting centers. In general form, information can be transmitted via both the wire and the wireless networks. In case of application of wireless networks, owing to the self-organization property and application of special protocols, it becomes possible to transmit the information by retransmission from one node to another using the transit nodes (T nodes).

V. Mochalov (✉) · M. Bersenev
Institute of Cosmophysical Research and Radio Wave Propagation FEB RAS, Mirnaya Str., 7, 684034 Kamchatka Region, Elizovskiy District, Paratunka, Russia
e-mail: sensorlife@mail.ru

M. Bersenev
e-mail: misha_589_bers@mail.ru

R. Cárdenas et al. (eds.), *Proceedings of the 2nd International Conference on BioGeoSciences*, https://doi.org/10.1007/978-3-030-04233-2_18

In paper [1], software–hardware complex «sensor signal analysis network» (SSAN) for construction and functioning of sensor networks for the distributed analysis of time-synchronized signals is considered. Under MSNERM, we will understand the sensor network, focused on electromagnetic radiation monitoring (ERM) over a wide frequency range and allowing the software to configure the functions performed by MSNERM by selecting the used antennas, ERM frequency bands, and software for signal analysis.

2 Hardware and Software Implementation of MSNERM F-Node

Figure 1 shows a generalized scheme of the MSNERM F-node, including a minicomputer, measurement time synchronization unit, ADC, sensors, communication modules, antenna switch, antennas, software-defined radio module, and optional field-programmable gate array (FPGA). In case of the absence of the necessity to use the FPGA, the ADC module is connected directly to the minicomputer. Depending on the conditions of application and accuracy requirements, the time synchronization unit may be realized on the basis of application of different types of satellite time synchronization GPS/GLONASS modules (NMEA 0183 signal comes to a mini-PC, PPS signal comes to ADC and FPGA input, and the stabilized output of 10 MHz from GPS/GLONASS module comes to FPGA input) and/or the connectives of temperature-stabilized precise highly reliable quartz oscillator with FPGA.

One possible implementation of the MSNERM F-node is shown in Fig. 2. Signals from sensors and VLF antennas come to preamplifiers and then to ADC (analog-to-digital converter). PPS signal from GLONASS/GPS module also comes

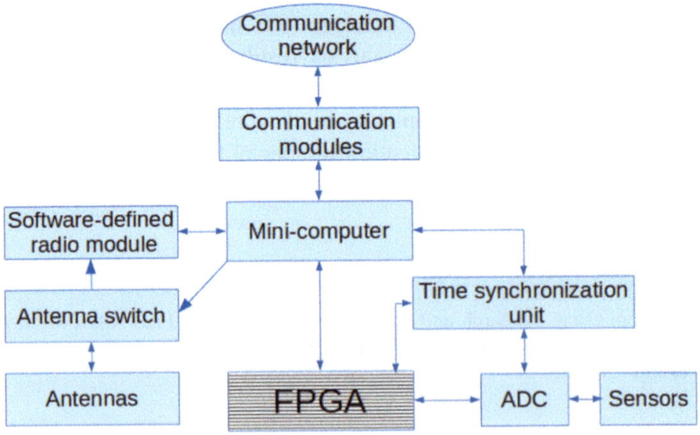

Fig. 1 Functional scheme of MSNERM F-node

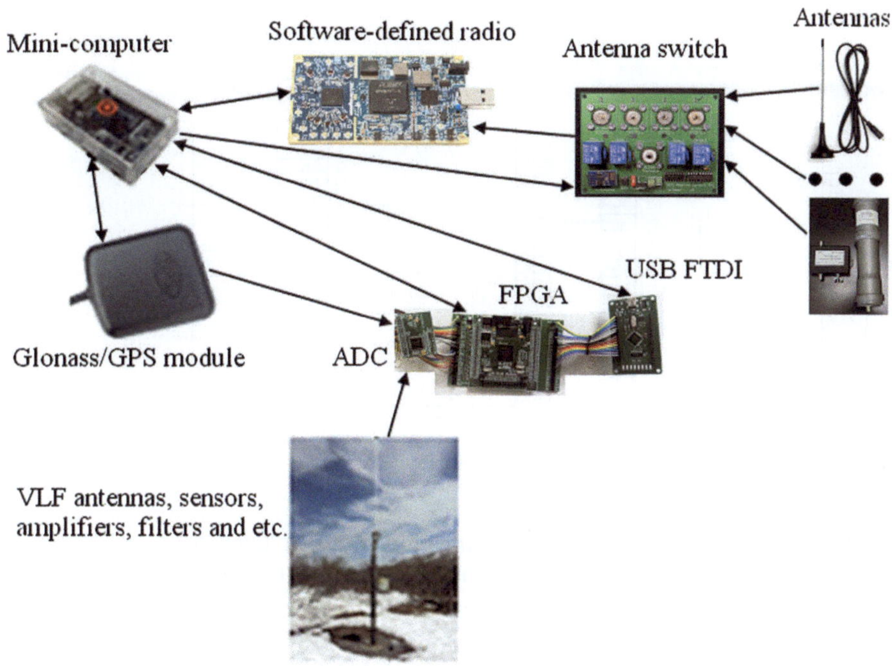

Fig. 2 One possible implementation of MSNERM F-node

to the ADC that allows us to make time synchronization of the measurements at different sensor network F-nodes. In paper [2], the process of stream identification of PPS-impulses using cubic spline interpolation is considered. NMEA 0183 signal comes to the minicomputer to set the system time and to determine the F-node location. After digitization, the general signal from the ADC input comes to the minicomputer for stream processing. Several programs are launched on the minicomputer that is running Linux operating system; they are: (1) program for data reading from ADC and (2) program of stream analysis (SA) of natural electromagnetic radiation sources based on the data registered in VLF range. Stream analysis program is capable of operating in two modes: (1) read mode of digitized data on different channels into a circular buffer inner structure and custom actions during every read record (almost real-time processing mode) and (2) record mode of files with a given duration with the following processing of files by custom actions. In paper [1], generalized functional scheme of signal stream analysis by SSAN F-node is considered.

Figure 3 shows the electrical circuit for connecting modules: ADC, FPGA, USB FTDI Bridge. The signal from antenna through the input filter and the amplifier is fed to the ADC. The ADC uses the HW-AD7606 module. The basis of this module is the Analog Devices AD7606 chip. The ADC has eight input channels and sampling rate up to 200 kHz with a resolution of 16 bits per channel [3].

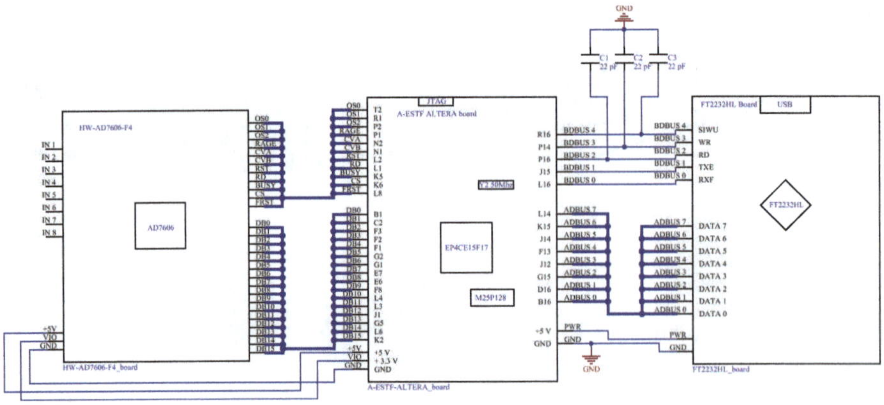

Fig. 3 An electrical circuit for connecting the module

Technical specifications of AD7606 chip are: (1) number of channels—8; (2) ADC resolution—16 bits; (3) input frequency 0–100 kHz; (4) sampling frequency—up to 200 kHz; (4) the input voltage ± 5 V, ± 10 V; (5) input resistance 1 MΩ; (6) voltage resolution range ± 5 V–152 μV, range ± 10 V–305 μV; and (7) the REF voltage—2.5 V. The operation of AD7606 is controlled by an Altera Cyclone IV EP4CE15 FPGA chip. This chip works in accordance with the firmware recorded in SPI flash memory chip M25P128, which polls the HW-AD7606-F4 according to the sampling rate. The firmware for FPGA is developed in the environment of Altera Quartus 2 Web Edition. The scheme of firmware EP4CE15 is shown in Fig. 4. Then, the received data from the HW-AD7606 module is transferred to the USB FTDI Bridge module for sending to the minicomputer via the USB interface. The USB bridge module is implemented on the FT2232HL chip and flash chip. For correct operation of FT2232HL chip (in 245 FIFO mode), it is necessary to reprogram the configuration flash chip. Reprogramming is performed using the utility FT_Prog provided on the manufacturer's site of the chip [4].

The broadband recording module consists of several antennas, an antenna switch, matching circuits, SDR transceiver LimeSDR-USB. The appearance of the LimeSDR-USB transceiver is presented in Fig. 5. SDR transceiver LimeSDR-USB has two RF inputs. This device is capable of receiving signals in the range of 100 kHz–3.8 GHz. The basis of this receiver is the LMS7002M transceiver chip, the Altera Cyclone IV EP4CE40F23 FPGA chip, the USB 3.0 Cypress, and USB 3.0 controller CYUSB3014-BZXC. One of the distinguishing features of this SDR is the open source code for firmware FPGA chips and USB controller, the presence of electrical circuits, which provides a wide range of options for adapting the product for specific tasks. Specifications received from the site are [4]: (1) operating frequency range: 100 kHz–3.8 GHz; (2) IF band: up to 61.44 MHz; (3) the resolution of the ADC—12 bits; and (4) the resolution of the DAC—12 bits.

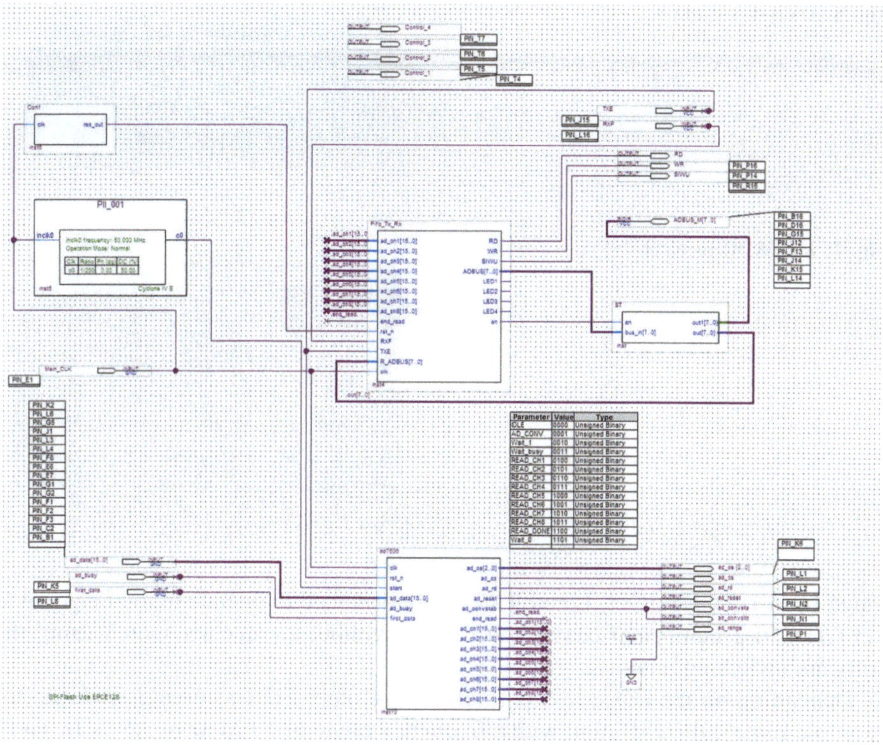

Fig. 4 Scheme of firmware EP4CE15 in the Quartus 2 editor

Fig. 5 SDR transceiver LimeSDR-USB

When using a wide frequency range from 100 kHz to 3.8 GHz, multiple antennas are required. An antenna switch with matching circuits is used to switch the antennas. In fact, LimeSDR-USB has six inputs: RX1_H, RX1_L, RX1_W, RX2_H, RX2_L, and RX2_W. RX1_H and RX2_H are designed to operate in the

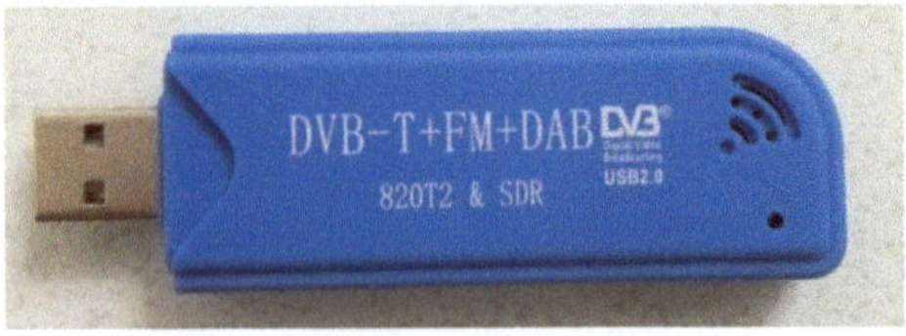

Fig. 6 DVB-T TV tuner

Fig. 7 RTL-SDR receiver

range from 1.5 GHz. RX1_L and RX2_L are designed to operate in the range up to 1.5 GHz. In a number of cases, you can do without an antenna switch using the software switching antennas.

Depending on the tasks, it is possible to use other SDR receiving modules, for example LimeSDR-USB mini with a frequency range of 10 MHz to 3.5 GHz, an IF band of 30.75 MHz [5], or RTL-SDR receiver. RTL-SDR receiver is based on the R820T and RTL2832U. This receiver is both in the form of a simple DVB-T TV tuner (Fig. 6) and in the form of a separate receiver SDR (Fig. 7).

Specifications of RTL-SDR receiver based on the R820T and RTL2832U are [6]: (1) frequency range: 24–1750 MHz; (2) IF band: from 0.25 to 3.2 MHz; and (3) the resolution of the ADC is 8 bits. In a number of descriptions as well as the RTL-SDR, there is also the possibility of receiving signals in the frequency range 0.1–24 MHz. This mode of operation is provided in two ways: (1) connection of the HF antenna input directly to the RTL2832U ADC. In this case, the amplification and filtering chains are absent. Such a receiver on this range has an extremely low sensitivity. (2) Application of the converter of the receiving band 0.1–24 MHz to the working band of the R820T chip is 24–1750 MHz.

3 Construction of MSNERM Structure

Construction of MSNERM requires the solution of many complicated problems from different areas of investigation, such as: investigation of radio physical characteristics of analyzed signals to select used antennas, development of network nodes, construction of physical and mathematical models for the electromagnetic radiation monitoring processes, construction and/or selection of a model of radio signal propagation, estimation of the effects of external processes on analyzed signals, development of the algorithms for recognition and automatic detection of analyzed signal characteristics, development of the methods and algorithms for MSNERM structure synthesis, estimation of measurement errors and limitations, consideration of spatial and other limitations for MSNERM nodes placement, estimation of the specified functional and structural parameters of the synthesized MSNERM, development of the functions to evaluate the MSNERM nodes placement.

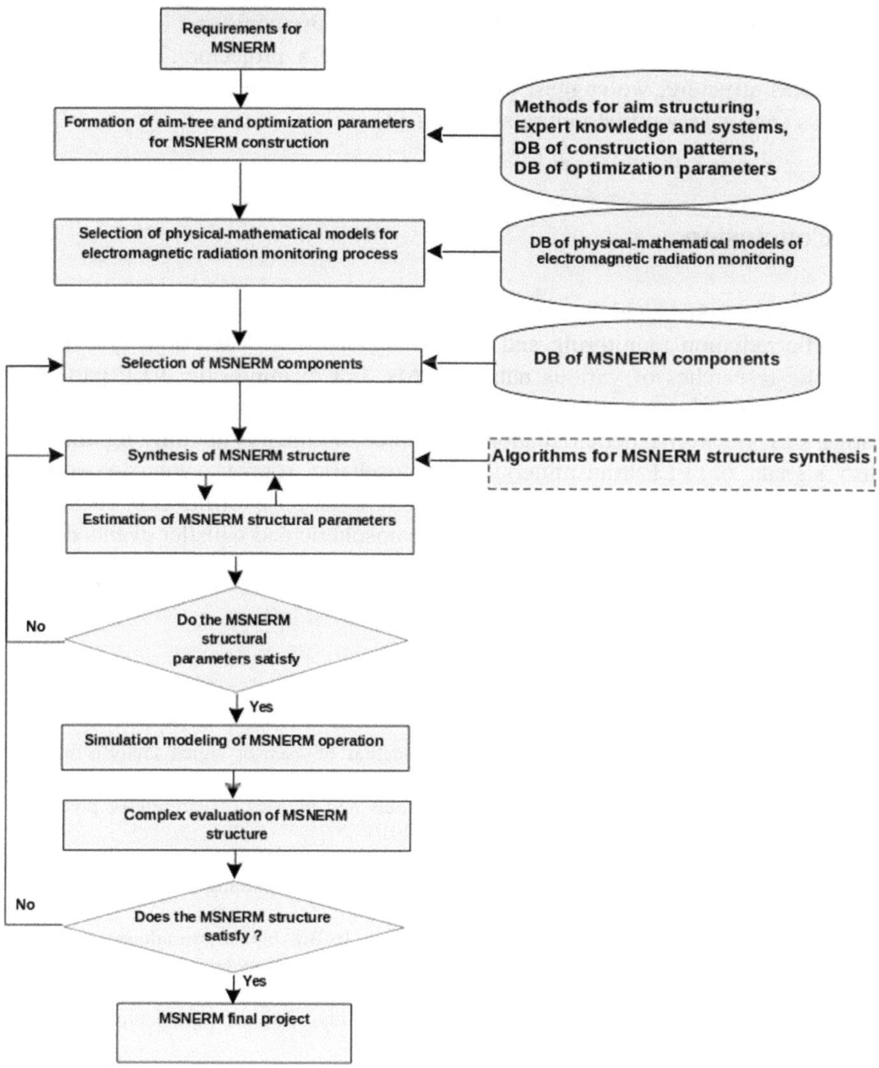

Fig. 8 Functional scheme of MSNERM construction

The initial data for MSNERM structure synthesis is the following: description of monitoring object (three-dimensional map of an object, its size, spatial, functional, and price limitation for node placement, electromagnetic state at the points available to place the MSNERM nodes, etc.), types of available nodes and their characteristics, applied physical and mathematical models of electromagnetic radiation monitoring, functional requirements, optimizing parameters, estimator of synthesized structures, and so on.

On the basis of [7–10] papers in Fig. 8, we propose the functional scheme for MSNERM construction. A functional block to synthesize the MSNERM structure is presented in Fig. 8. The main task of it is the placement of MSNERM nodes so that the synthesized network would satisfy the requirements and aims of the projection, i.e., would have "desirable properties" of a projector. To synthesize a MSNERM structure, we suggest to apply a multi-agent bio-inspired algorithm of wireless sensor network structure synthesis [7].

4 Conclusion

The complex considered in the work can be used for various tasks for electromagnetic radiation monitoring and performing distributed and time synchronized scientific researches of various nature ERMs. For example, the VLF part of the considered complex was used to construct a distributed sensor network for monitoring VLF electromagnetic radiation, the source of which is not only lightning, but also signals of VLF transmitters, magnetospheric sources, volcano explosive eruptions, tropical cyclones, and much more. On the basis of the constructed VLF sensor network, we created a database for atmospheric and whistler events detected in the Russian Far East [11].

References

1. Mochalov VA, Mochalova AV (2017) Application of «sensor signal analysis network» complex for distributed, time synchronized analysis of electromagnetic radiation. In: viii international conference "solar-terrestrial relations and physics of earthquake precursors", pp 6. https://doi.org/10.1051/e3sconf/20172002010
2. Mochalov V, Sannikov D, Karimov R, Shevtsov B, Drugin G, Cherneva N, Mochalova A, Lichtenberger J, Argunov V (2016) VLF sensors for lightning research. Procia Eng 168:1721–1724
3. Datasheets of AD7606 8-/6-/4-Channel DAS with 16-Bit, bipolar, simultaneous sampling ADC. http://www.analog.com/media/en/technical-documentation/data-sheets/AD7606_7606-6_7606-4.pdf
4. FT_PROG 3.3.88.402-EEPROM programming utility. http://www.ftdichip.com/Support/Utilities.htm#FT_PROG
5. LimeSDR. https://myriadrf.org/projects/limesdr/

6. Datasheets of R820T. https://www.rtl-sdr.com/wp-content/uploads/2013/04/R820T_datasheet-Non_R-20111130_unlocked1.pdf
7. Mochalov VA (2015) Multi-agent bio-inspired algorithms for wireless sensor network design. In: Proceeding of the IEEE 17th international conference on advanced communication technology, ICACT 2015, Phoenix Park, Korea, pp 34–42
8. Mochalov VA (2015) Synthesis of the wireless sensor network structure in the presence of physical attacks. Lect Notes Comput Sci 9247:11–22
9. Mochalov VA, Mochalova AV (2017) Algorithms for changing the structure of geospace self-organizing question-answering sensor networks. In: VIII international conference "solar-terrestrial relations and physics of earthquake precursors", pp 11. https://doi.org/10.1051/e3sconf/20172002009
10. Mochalov VA, Mochalova AV, Chowkwale B, Homutov SU (2016) Functional scheme of the environment monitoring systems architecture design. In: Proceedings of the IEEE 18th international conference on advanced communication technology, ICACT 2016, Phoenix Park, Korea, pp 289–295. https://doi.org/10.1109/icact.2016.7423364
11. Mochalov VA, Drugin GI, Karimov RR, Shevtsov BM, Cherneva NV, Mochalova AV, Permyakov MS, Droga AN (2018) Creation of a database for atmospheric and whistler events detected in the Russian Far East. VarSITI Newslett 16:8–9

Tilt and Orientation of a Flat Solar Collector to Capture Optimal Solar Irradiation in Chilean Latitudes

Lisdelys González-Rodríguez, Laura Pérez, Adelqui Fissore, Lien Rodríguez-López and Jorge Jimenez

Abstract The use of solar technologies is growing steadily throughout the world since solar radiation is recognized as an inexpensive and local renewable source of energy. At the same time, it helps to reduce the great environmental impact caused by the exploitation of non-renewable energy sources. Accurate information about incident solar radiation over an inclined surface is important for selection and installation of solar technologies. In this study, seasonal and annual total irradiation values received on a flat solar collector were calculated by varying the inclination angle from 0° to 90°, and the azimuth angle from −180° to 180° for 12 cities in Chile, based in one typical meteorological year. The study provides useful information about the influence of the tilt and azimuth angles to the total seasonal and annual solar energy collected. Several cities of Chile located from north to south were selected for the study, including Arica, Calama, Antofagasta, Vallenar, La Serena, Valparaíso, Talca, Concepción, Los Angeles, Valdivia, Puerto Montt, and Puerto Natales. The isotropic sky model was used to calculate the total irradiances. The results showed that during the summer months the average irradiation values were above 150 kWh/m^2 in all cities. During the summer months, the device may be inclined between 0° and 30°, whereas during the winter months it is recommended to reset the inclination angle between 30° and 60°. The estimated annual solar radiation gains, based on tilt and azimuth angles, compared to a horizontal surface, increase toward the south with a maximum gain 10.08% for the city of Puerto Natales. The lowest gain was for the city of Arica with 0.55%, which shows that its best use is close to the horizontal surface. Although, the range of inclination and azimuth angles to achieve optimal irradiation values is wide, it is recommended to set the tilt angle of the flat solar collector between $0° \leq \beta \leq 60°$ and the azimuth angle from $-60° \leq \gamma \leq 60°$. For losses smaller than 5% of irradiation, the azimuth angle can oscillate between ±30° without a significant impact on the total irradiation captured by a flat solar collector.

L. González-Rodríguez (✉) · L. Pérez · A. Fissore · J. Jimenez
Faculty of Engineering, University of Concepcion, Concepcion, Chile
e-mail: lisdegonzalez@udec.cl

L. Rodríguez-López
Environmental Science Centre, EULA, University of Concepcion, Concepcion, Chile

R. Cárdenas et al. (eds.), *Proceedings of the 2nd International Conference on BioGeoSciences*, https://doi.org/10.1007/978-3-030-04233-2_19

Keywords Solar irradiation · Renewable energy · Tilt angle · Flat solar collector Chile

1 Introduction

One of the main challenges of sustainable development, both globally and nationally, lies in responding to critical issues related to energy sources and utilization. There is an urgent need to implement renewable energy sources that replace fossil fuels. In this sense, the most abundant energy resource is solar energy, which would cover all energy needs if it were to increase its efficiency of use, becoming an inexpensive source, renewable, sustainable, and non-polluting energy. This solar radiation potential can be used in various applications such as desalination, heating, photocatalytic reactors, photovoltaic panels, and thermal solar collectors. It is known that the proper installation of these technologies can have a notable change in the expected performance, since the climatic conditions, the latitude, the altitude, the orientation, and the angle of inclination of the collecting device may affect the performance of the technology in terms of total energy yields.

On the other hand, due to the scarce meteorological data and the high cost of instruments for measurements of irradiance, several authors have developed techniques and models that provide a satisfactory approximation of the behavior of solar radiation. Among the reference techniques are the isotropic sky [1] and anisotropic [2] models. These models and techniques have been used in several studies in order to optimize the orientation and inclination of the solar collector and thus achieve maximum irradiation capture. In the area of Ma'an, Egypt [3], it was showed that up to 16.1% of extra radiation can be collected with an annual fixed inclination angle of 28.7° and with an orientation to the south. The authors also suggest that solar panels should be adjusted at least four times a year to increase the collection capabilities. Another study in western Ghana [4] reported that for solar collectors, optimum tilt angles should vary between 16° and 20°. In the periods January–March and September–December, the angle had a value equal to the latitude plus 16º, while for the period from April to August the optimum inclination was the latitude of the location plus 20º. On the other hand, Jafarkazemi and Saadabadi [5] applied the k_T method to evaluate the consequences of orientation in the configuration of the solar collectors, reported that the optimal annual inclination angle was 22°, very close to the latitude of Abu Dhabi, and the orientation angle was to the south direction. Another study in the high Andean Equatorial Zone concluded that in order to optimize the incidence of direct solar radiation for water heating with vacuum tubes, an inclination angle equal to horizontal should be used [6].

Chile, for its geographical conditions, is an interesting benchmark for researching technologies on solar energy. Chile has good solar radiation during the whole year with solar radiation levels that exceed 2500 kWh/m^2 in some areas of the north [7], with clear skies year-round. Most of the designs of systems for harnessing solar radiation in the southern hemisphere include modules that points to the north with a fixed azimuth angle equal to the local latitude plus 10°, which allows to maximize the generation of energy during winter months. However, this is not optimal for all locations, since the angle of inclination and azimuth may vary in some cases. A study conducted in the center of Chile, regarding the orientation of photovoltaic panels for an economic maximization in generation plants and their coupling with the demand profile found for that for a fixed orientation the power facility would have optimum annual economic performance for panels facing the north [8]. However, during the summer months (November–January), the economic income for a photovoltaic plant is optimized by orienting the panels to the west (270°) and northwest (280°–290°). Another study [9] regarding the total annual solar radiation received on 701 roof plans in Valparaíso shows that the ceilings that receive the highest annual solar radiation are oriented toward the north and the northwest and the lowest solar intensities are obtained in the oriented roofs to the south and southeast. However, the total annual solar radiation can be greatly influenced by the angle of inclination of the roof. Therefore, in some roof planes facing south with an angle of inclination less than 15° can receive higher solar intensities than others facing north. The foregoing demonstrates the importance of understanding the impact of the inclination and azimuth angle of a flat solar device in order to the maximum energy solar captain in different location of Chile. The present study intends then to promote the application of renewable solar energy in Chilean cities provides useful information for assessing and implementing technologies for solar energy utilization, such as: photovoltaic panels, solar photocatalysis (orientation of photocatalytic reactors), sanitary hot water systems, solar cooker and agriculture (fruit drying) among others.

2 Methodology

2.1 Study Area

Chile is located on the southwestern margin of South America, between 17° 30′ and 56° 30′ south latitude. It comprises 15 regions including the Metropolitan Region where the capital, Santiago, is located. The country has a wide variability of climates, due to the topographic features, the influence of the ocean and the latitudinal amplitude that covers almost 40° degrees [10]. Throughout the year, the territory receives a good amount of solar radiation, especially in the northern regions. For this study, 12 Chilean cities were selected for analyses, which included populated areas with industrial and economic activities with different energy needs in terms of access, costs, and demand. The criteria for selecting the locations were based on adequate spacing to cover a range of latitudes, as well as the availability of climatological data in that area. The cities under study, from north to south, are Arica, Calama, Antofagasta, Vallenar, La Serena, Valparaiso, Talca, Concepcion, Los Angeles, Valdivia, Puerto Montt, and Puerto Natales, as shown in Fig. 1. The selected cities captured the climate variety throughout the territory, from north to south and from the coast to the Andean mountain range.

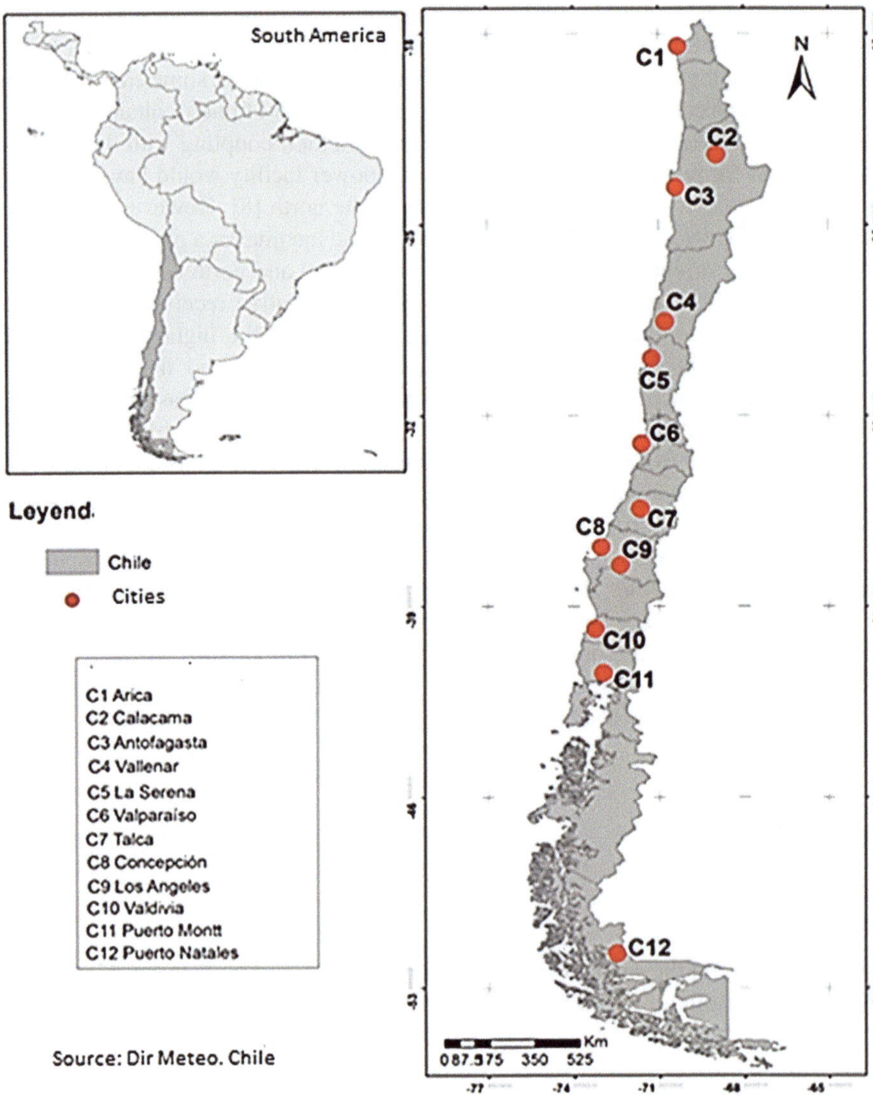

Fig. 1 Location of the 12 cities selected in Chile

2.2 *Calculation Algorithm*

From the Meteonorm v6.1 [11], the global radiation values (W/m^2) were obtained, during a typical year, for each of the selected cities. The radiation data were validated with local values of monthly averages obtained from the Chilean Meteorological Directorate (www.meteochile.cl). These data are normally recorded for a horizontal plane, without distinguishing direct component (Ib) and diffuse (Id).

The transformation to the inclined plane is done through geometric relationships and factors that require knowing where the radiation comes from. For this analysis, it was necessary to include several angles that describe the movement and position of the Sun. While the first, three corrections are made to convert the local time to the solar time; the correction of length, time, and summer time. This last correction is made once in summer to obtain better use of sunlight and energy savings. The declination of the Sun (δ) is determined according to [2] and is calculated by the following Eq. 1:

$$\delta = 23.45\,\mathrm{sen}(360(284 + n)/365) \tag{1}$$

where n represents the correlative day of the year (1–365) as of January 1. This angle is measured for a reference system with the z-axis on the north–south axis. Due to the Earth's own inclination, this angle has a value of 23.45° in winter solstice and −23.45° in summer solstice. The general relationship between the parameters of declination, latitude (ϕ), slope or slope (β), surface azimuth angle (γ), solar time angle (ω), and the angle of incidence of the radiation beam (θ) is defined by Eq. 2:

$$\begin{aligned}\cos\theta = {} & \mathrm{sen}\delta\mathrm{sen}\phi\cos\beta - \mathrm{sen}\delta\cos\phi\mathrm{sen}\beta\cos\gamma + \cos\delta\cos\phi\cos\beta\cos\omega \\ & + \cos\delta\mathrm{sen}\phi\mathrm{sen}\beta\cos\gamma\cos\omega + \cos\delta\mathrm{sen}\beta\mathrm{sen}\gamma\mathrm{sen}\omega\end{aligned} \tag{2}$$

On the other hand, to take into account the inclination of the receiving surface, the term (Rb) is calculated from the following Eq. 3:

$$\mathrm{Rb} = \cos\theta/\cos\theta z \tag{3}$$

Where: θz is the angle of the zenith and is obtained from the subtraction between 90 and the angle of altitude. In this way, the values of Rb are obtained and the maximum possible radiation is calculated on a horizontal surface in the absence of atmosphere (I_0), which is deduced by Eq. 4:

$$I_0 = \mathrm{Gsc}\ [1 + 0.033\cos(360n/365)](\mathrm{sen}\phi\mathrm{sen}\delta + \cos\phi\cos\omega\cos\delta) \tag{4}$$

In Eq. 4, the term G_{SC} is the solar constant at the top of the atmosphere with a value equal to 1367 W/m^2. For the calculation of the diffuse radiation fraction (Id), the Orgill and Hollands correlation is used [12], but a preliminary calculation of the atmospheric transparency index (kT) is performed beforehand. This index is the quotient between the global irradiation (I) and the incident power on a horizontal plane (I_0) at the considered latitude [2] as can be seen in Eq. 5:

$$\mathrm{kT} = I/I_0 \tag{5}$$

This index is considered the attenuation factor of the atmosphere. In general, when the atmosphere is clearer, a smaller fraction of radiation disperses. In addition, the clarity index is a stochastic parameter, which is a function of the time of year, season, weather conditions, and geographical location. This correlation has been used, since it is based on time. Finally, the sky model used for the calculations of the total irradiation (I_T) for different inclinations and orientations is the isotropic sky model according to Eq. 6:

$$I_T = \text{IbRb} + ((1 + \cos\beta)/2)\text{Id} + ((1 - \cos\beta)/2)(\text{Ib} + \text{Id})\rho \quad (6)$$

The reflectance value (ρ) is taken for all cases equal to 0.2; this value is estimated for earthy surfaces. In all cases, starting from daily hourly values, the monthly and annual I_T calculation is performed. Different orientations ($-180° \leq \gamma \leq 180°$) and inclination angles ($0° \leq \beta \leq 90°$) are considered at intervals of 15° each. To represent the southern hemisphere, the 0° orientation is selected as north, and positive values are taken for the west and negatives toward the east. In total, I_T is calculated for 12 cities in Chile, with 7 inclinations and 24 different orientations; all for each month of a typical year. Different units are used to express the value of the insolation of a place; the most convenient for our study is the Kilowatt hour per square meter (kWh/m^2).

2.3 *Software*

Various tools are used as support, among them the well-known reference for the generation of meteorological data, based on more than 25 years of experience in the development of Meteonorm v6.1 meteorological databases for energy applications. In the construction of maps and location of cities, the geographical information systems software ArcGIS v10.1 is used and Python v3.6 is used to create contour graphics.

3 Results and Discussion

The behavior of the irradiation on the Earth's surface varies considerably during the year since it is determined by the inclination of the surface with respect to the Sun's rays. This depends essentially on the day, time, altitude, and latitude of each location (city). Previously, a general analysis is made of the monthly behavior of the I_T values on the horizontal plane for the selected cities of the Chilean territory, which represented in Fig. 2.

The results obtained were favorable and according to what was expected. In general, the lowest monthly irradiation behavior is between the winter months, mainly June and July. In contrast, during the summer season, mainly December and January, for all the cities studied, the irradiation values increase until reaching values above 150 kWh/m^2. Throughout the year, the best cumulative total monthly

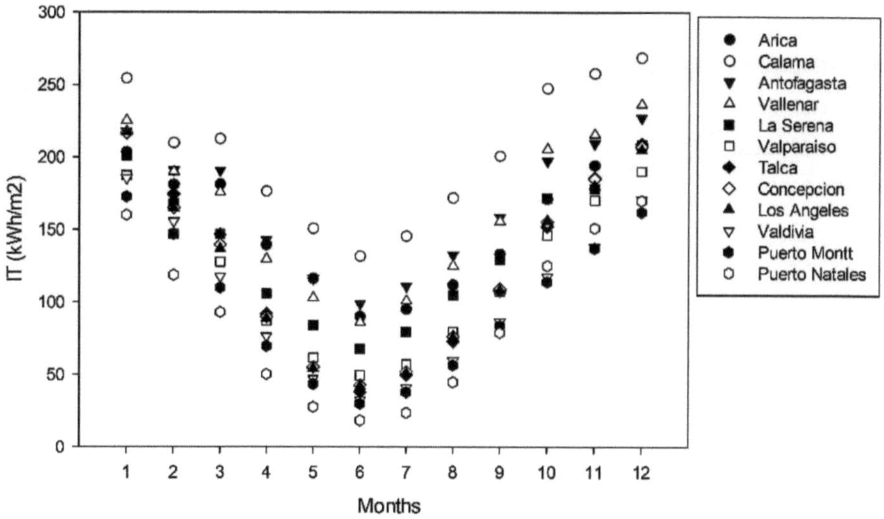

Fig. 2 Total monthly irradiation on a horizontal surface for cities in Chile

I_T is for the city of Calama (North of the territory), with a maximum value in December of 269 kWh/m^2 and a minimum in June with 132 kWh/m^2. Other cities located in the north, such as Antofagasta, Vallenar, and Arica obtained I_T values of 227, 237, and 206 kWh/m^2, respectively. As the latitude increases to the south, the monthly total I_T decreases progressively until reaching a minimum of 24 kWh/m^2 in July for the city of Puerto Natales (southern most city). It is evident that the seasons of the year determine the values of radiation on the terrestrial surface. To analyze how the angle of inclination of a solar collector affects the seasonal I_T values at each location, calculation was made maintaining the orientation toward the north and varying the inclination 15° at a time. Figure 3 shows the results obtained for each location.

Figure 3 shows that the irradiation values on a flat surface increase by varying the horizontal position (0°) to an appropriate angle of inclination, which depends on the latitude and the day of the year in which it is located, and then reduce gradually as the inclination continues to increase. Thus, optimal solar radiation values are achieved for each month by varying the angle of inclination within a recommended range. Generally, to obtain maximum monthly irradiation values throughout the year, the inclination should be between $0° \leq \beta \leq 30°$. Seasonally, in the months of summer and spring, the best accumulative irradiation is obtained with tilt angles between $0° \leq \beta \leq 30°$. On the contrary, in winter and autumn months, the maximum irradiation values are achieved by increasing the inclination toward values between $30° \leq \beta \leq 60°$.

Among the cities studied, the cities of Calama and Puerto Natales constitute the extreme behaviors, due to their maximum and minimum irradiation values throughout the year. In Calama, the best radiation accumulations year-round are

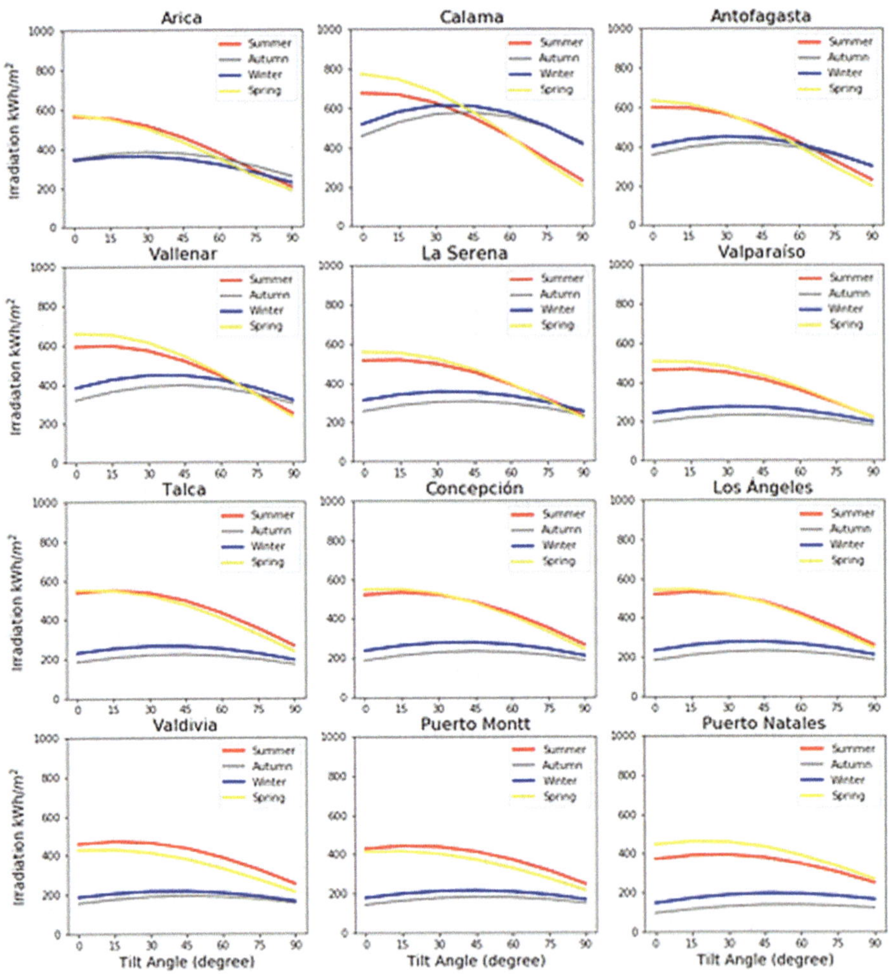

Fig. 3 Total seasonal irradiation by varying the tilt angle for cities in Chile

obtained with a tilt angle equal near 15°, and it is important to note that in the colder seasons (winter and autumn) the irradiation values exceed up to 48% those obtained in the warmer seasons for angles greater than 45°.

On the other hand, the Puerto Natales city, representative of the southern Chilean territory, has a similar behavior in the different seasons of the year when the inclination varies. The irradiation in spring and summer is much higher than in the rest of the year for all angles of inclination. From Fig. 3, it is concluded that solar radiation is an intensive function of the angle of inclination of the surface.

Table 1 shows the maximum monthly irradiation value obtained with the optimum tilt angle for each of the cities and seasons of the year.

Table 1 Maximum monthly irradiation values (kWh/m^2) for optimal tilt angles in different cities

Cities	Summer	Autumn	Winter	Spring
Arica	565 (0°)	382 (30°)	358 (30°)	572 (0°)
Calama	677 (0°)	578 (45°)	613 (30°)	774 (0°)
Antofagasta	599 (0°)	417 (45°)	451 (30°)	634 (0°)
Vallenar	595 (15°)	394 (45°)	444 (30°)	658 (0°)
La Serena	519 (15°)	307 (45°)	356 (30°)	559 (0°)
Valparaiso	466 (15°)	234 (45°)	275 (30°)	507 (0°)
Talca	549 (15°)	219 (45°)	264 (30°)	547 (0°)
Concepcion	531 (15°)	232 (45°)	278 (30°)	543 (15°)
Los Angeles	471 (15°)	192 (45°)	216 (45°)	428 (15°)
Valdivia	471 (15°)	192 (45°)	216 (45°)	428 (15°)
Puerto Montt	444 (15°)	182 (45°)	215 (45°)	416 (15°)
Puerto Natales	392 (30°)	138 (60°)	196 (45°)	462 (15°)

This information is of great help, since it shows in the year and seasonally how many times the inclination of the flat solar collector must be adjusted if the geometry of the design allows it. On a seasonal basis, this optimum tilt angle increases slightly in the winter and autumn months and decreases to the minimum value in the summer and spring months due to the change in the Sun's position. The cities such as: Arica, Los Angeles, Valdivia, and Puerto Montt must adjust the inclination of the solar collector three times a year. In Arica city, the maximum irradiation is obtained with inclination of 0° in the warmer months (summer and spring) and for the cold months is reached with inclination of 30°. On the other hand, Los Angeles, Valdivia and Puerto Montt shows the maximum irradiation for the warmer months with inclination of 15°, while for the other seasons the maximum is obtained with 45°. The rest of the cities could be adjusted four times in the year, Calama and Antofagasta in the summer and spring reach its maximum irradiation with an angle equal to the horizontal and for autumn and winter is reached with 45° and 30°, respectively. Vallenar, La Serena, Valparaiso and Talca cities present its maximum irradiation in summer with tilt angle equal 15°, go through by 45° in the autumn, 30° in winter and 0° in Spring. The city of Concepcion reaches the maximum values of seasonal irradiation with tilt angles equal to the cities previously mentioned, with the exception of spring, where the appropriate tilt angle would be 15°. Puerto Natales is the last city studied; its maximum irradiation in the summer is obtained with tilt angle of 30°, while in the seasons of autumn, spring, and winter is obtained with 60°, 45°, and 15°, respectively. Even for cities far south, such as, Puerto Montt and Puerto Natales, significant irradiance values are obtained with tilt angles of $\beta = 90°$ for the cold months.

Subsequently, in order to know which working conditions would be the most appropriate to take advantage of the greater amount of irradiance received, the annual irradiation totals on the solar collectors are analyzed, varying the inclination and the orientation of the collecting surface for all the cities, which gives a complete and visual representation of the fundamental interest in this report. Figure 4 shows the relative values of irradiation as a function of the totals on a horizontal surface, by means of contour graphs for each city.

In Fig. 4, it is observed that for all the cities, not only varying the angle of inclination of the surface but correcting the orientation or azimuth, a percent of use of the greater irradiation is obtained. It can also be observed that the maximum annual total I_T values are within an inclined surface between 15° and 30°. However, in general, the range for which the combination of tilt/azimuth is recommended in obtaining optimum values of $\geq 95\%$ with respect to the horizontal is ample: It slopes between $0° \leq \beta \leq 60°$, with azimuth from $-60° \leq \gamma \leq 60°$. However, the azimuth maintains a distribution clearly directed toward the north.

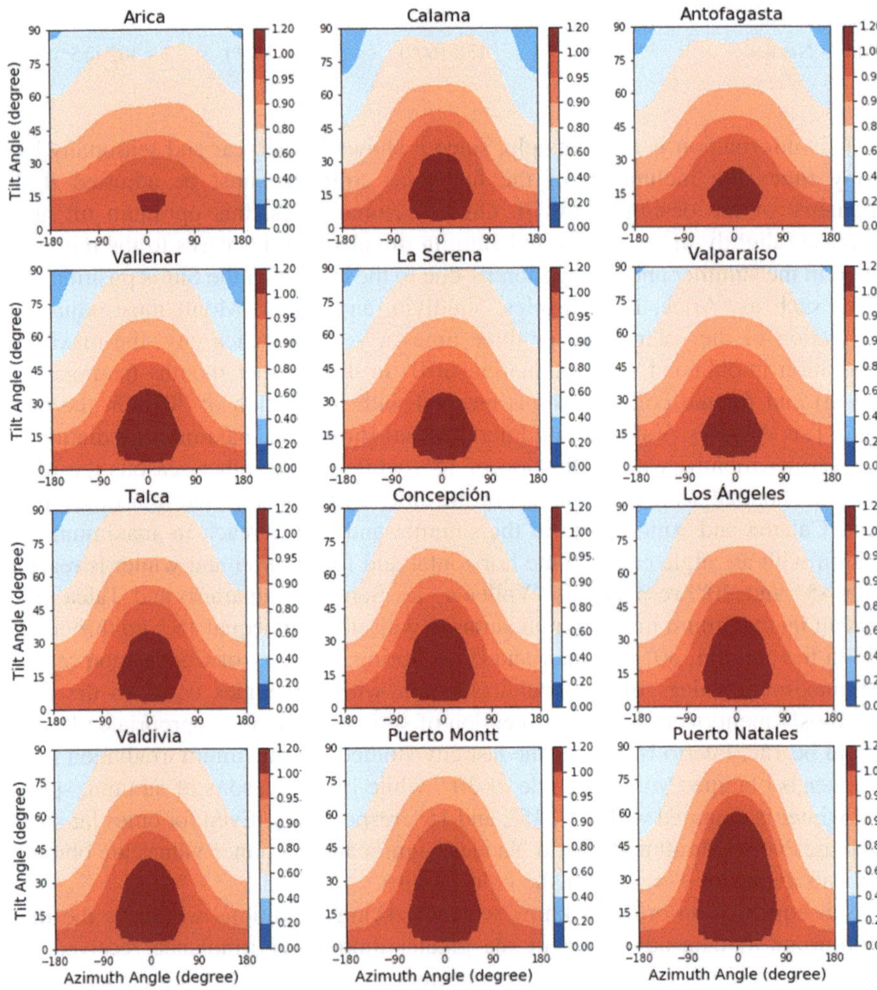

Fig. 4 Total irradiation relative to the horizontal for different tilt and azimuth angles

Table 2 Maximum annual irradiation on a plane inclined to different azimuth and tilt angles, with respect to the horizontal and percentage gain of the inclined plane compared to that received on a horizontal plane for different cities in Chile

Cities	Latitude (ϕ) and Altitude (mals)	Horizontal I_T (kWh/m^2 year)	Maximum I_T for a given tilt angle (kWh/m^2 year)	Maximum I_T varying the optimum tilt angle seasonally (kWh/m^2 year)	Gain annual I_T to the horizontal (%)	Gain annual I_T using the seasonal optimum angle	Tilt range (β) for > 95% irradiation	Azimuth range (γ) for > 95% irradiation
Arica	18.47°, 2	1823	1833 ($\beta = 15°$)	1877	0.55	2.88	0°–30°	−45° to 45°
Calama	22.45°, 2400	2428	2527 ($\beta = 15°$)	2642	4.08	8.10	0°–30°	−60° to 60°
Antofagasta	23.64°, 40	1993	2047 ($\beta = 15°$)	2101	2.71	5.14	0°–30°	−60° to 60°
Vallenar	28.57°, 380	1948	2031 ($\beta = 15°$)	2091	4.26	6.84	0°–45°	−30° to 30°
La Serena	29.91°, 50	1644	1702 ($\beta = 15°$)	1740	3.53	5.52	0°–45°	−30° to 30°
Valparaiso	33.05°, 10	1410	1455 ($\beta = 15°$)	1481	3.19	4.79	0°–45°	−30° to 30°
Talca	35.43°, 102	1500	1555 ($\beta = 15°$)	1582	3.67	5.18	0°–45°	−30° to 30°
Concepcion	36.82°, 12	1494	1559 ($\beta = 15°$)	1596	4.35	6.39	0°–45°	−45° to 45°
Los Ángeles	37.50°, 133	1479	1544 ($\beta = 15°$)	1583	4.39	6.57	0°–45°	−45° to 45°

(continued)

Table 2 (continued)

Cities	Latitude (ϕ) and Altitude (mals)	Horizontal I_T (kWh/m^2 year)	Maximum I_T for a given tilt angle (kWh/m^2 year)	Maximum I_T varying the optimum tilt angle seasonally (kWh/m^2 year)	Gain annual I_T to the horizontal (%)	Gain annual I_T using the seasonal optimum angle	Tilt range (β) for > 95% irradiation	Azimuth range (γ) for > 95% irradiation
Valdivia	39.83°, 5	1227	1280 (β = 15°)	1308	4.32	6.19	0°–45°	−45° to 45°
Puerto Montt	41.46°, 14	1163	1230 (β = 30°)	1258	5.76	7.55	0°–45°	−45° to 45°
Puerto Natales	51.73°, 3	1061	1168 (β = 30°)	1188	10.08	10.69	0°–60°	−45° to 45°

Note Interpretation of columns as above description: name of the city; local latitude and altitude; incident annual irradiation on a horizontal surface; annual maximum irradiation at different tilt whit azimuth north; annual maximum irradiation varying tilt angle seasonally (three to four times depending on the city); maximum percentage of radiation gained respect horizontal; tilt and azimuth range for irradiation greater than 95%; as obtained from using the methodology described

Table 2 lists the selected cities by recording their latitude, altitude, and the cumulative average annual I_T values on a horizontal plane, in addition to the maximum annual irradiation that would be received with the optimum fixed angle of inclination and maximum annual irradiation when changing the angle of inclination for different season. Columns 6 and 7 shows the percentage gain of irradiation obtained by placing the solar collector at the optimum angle and the percentage gain of irradiation obtained when change for season the tilt angle. To know the use of the resource above 95%, the recommended range of values for tilt and azimuth angles is located in the last columns.

The maximum annual irradiation values are obtained in Calama, the city with the highest altitude studied, which reaches the value of 2527 kWh/m^2 year, followed by Antofagasta and Vallenar with 2047 and 2031 kWh/m^2 year, respectively. While in Puerto Natales, the irradiation reached 1168 kWh/m^2 per year.

The estimated annual solar radiation gains, based on tilt and azimuth angles, compared to a horizontal surface, increase toward the south with a maximum gain (10.08%) for the city of Puerto Natales. The lowest gain was for the city of Arica with 0.55%, which shows that its best use is close to the horizontal surface.

Even an improved percentage increase could be achieved by varying the angle of inclination per season of the year. However, the profits would be 2–3% in most cities with respect to the horizontal, exceeding the city of Calama that would earn 4% more than the annual irradiation, going from 4.08 to 8.10%. Therefore, a large economic investment in equipment to change the seasonal inclination angle in the flat solar collector would not be inappropriate.

With respect to the working range of the angle of inclination, this presents an increase toward the south. In Arica, recommended inclination angles are from 0° to 30°, while for Puerto Natales the range is between 0° and 60°. Finally, for losses smaller than 5% of irradiation, the azimuth angle can oscillate between ±30° without a significant impact on the total irradiation captured by a solar device.

4 Conclusions

The study allows to know the behavior of the solar resource by varying the conditions of inclination and orientation of a flat solar collector, which guarantees an appropriated installation of these solar technologies. Solar energy obtained from a flat collector can be improved by making adjustments of the surface inclination. This may improve the use of this local clean energy resource in different cities of Chile.

The irradiation values vary clearly by making adjustments in the tilt and azimuth angles of the solar devices. Seasonally, in the months of summer and spring, the best accumulative irradiation is obtained with tilt angles between $0° \leq \beta \leq 30°$. On the contrary, in winter and autumn months, the maximum irradiation values are achieved by increasing the inclination toward values between $30° \leq \beta \leq 60°$

One combination more general of tilt/azimuth is obtained; the inclination values can be oscillated between 0°and 60° and orientation values from $-60° \leq \gamma \leq 60°$ to obtain optimal irradiation values greater than 95% with respect to the horizontal.

Acknowledgements The authors want to thank National Commission for Scientific and Technological Research in Chile (CONICYT) for PhD grants.

References

1. Liu B, Jordan R Daily insolation on surfaces tilted towards equator. United States.
2. Reindl DT, Beckman WA, Duffie JA (1990) Evaluation of hourly tilted surface radiation models. Sol Energy 45(1):9–17
3. Altarawneh IS, Rawadieh SI, Tarawneh MS, Alrowwad SM, Rimawi F (2016) Optimal tilt angle trajectory for maximizing solar energy potential in Ma'an area in Jordan. J Renew Sustain Energy 8
4. Uba FA, Sarsah EA (2013) Optimization of tilt angle for solar collectors in WA, Ghana. Pelagia Res Libr Adv Appl Sci Res 4:108–114. www.pelagiaresearchlibrary.com
5. Jafarkazemi F, Saadabadi SA (2013) Optimum tilt angle and orientation of solar surfaces in Abu Dhabi. Renew Energy 56. https://doi.org/10.1016/j.renene.2012.10.036
6. Recalde C, Cisneros C, Ávila C, Urquizo G (2015) Influencia del ángulo de inclinación de los tubos solares evacuados sobre la temperatura del agua. Inf Tecnol 26:89–96. https://doi.org/10.4067/S0718-07642015000400012
7. Cornejo L, Martín-Pomares L, Alarcón D, Blanco J, Polo J (2017) A through analysis of solar irradiation measurements in the region of Arica Parinacota, Chile. Renew Energ, 112, 197-208.
8. Elton M, Lofat M, Sinclair CMS (2015) Estudio respecto a la orientación de paneles fotovoltaicos para una maximización económica de plantas de generación y su acoplamiento con el perfil de demanda. Dtsch Gesellschaft Für Int Zusammenarbeit GmbH 1:18
9. Araya-Muñoz D, Carvajal D, Sáez-Carreño A, Bensaid S, Soto-Márquez E (2014) Assessing the solar potential of roofs in Valparaíso (Chile). Energy Build 69:62–73. https://doi.org/10.1016/j.enbuild.2013.10.014
10. Gobierno de Chile (2018) Biblioteca del Congreso Nacional de Chile BCN. Inf Territ (n.d.). https://www.bcn.cl/siit/nuestropais/index_html. Accessed on 26 Jan 2018
11. Meteonorm Irradiation data for every place on Earth. http://www.meteonorm.com/
12. Orgill, Hollands (1977) Correlation equation for hourly diffuse radiation on a horizontal surface. Sol Energy 19(4):357–359

MIX
Papier aus verantwortungsvollen Quellen
Paper from responsible sources
FSC® C105338

If you have any concerns about our products,
you can contact us on
ProductSafety@springernature.com

In case Publisher is established outside the EU,
the EU authorized representative is:
Springer Nature Customer Service Center GmbH
Europaplatz 3, 69115 Heidelberg, Germany

Printed by Libri Plureos GmbH
in Hamburg, Germany